Resettlement Policy in Large Development Projects

Hydropower generation by construction of large dams attracts considerable attention as a feasible renewable energy source to meet the power demand in Asian cities. However, large development projects cause involuntary resettlement. Of the world's 40 to 80 million resettlers, many resettlers have been unable to rebuild their livelihoods after relocation and have become impoverished.

Resettlement Policy in Large Development Projects uniquely explores the long-term impacts of displacement and resettlement. It shows that long-term post-project evaluation is necessary to assess the rehabilitation and livelihood reconstruction of resettlers after relocation. It focuses on large dam projects in a number of Asian countries, including Indonesia, Japan, Laos, Turkey, Sri Lanka, and Vietnam, which are often ignored in displacement studies in favour of China or India. Drawing on a wealth of empirical data gathered over ten years, it presents crucial factors for successful resettlement by analyzing lessons learned. The range of countries allows for a diverse and complex set of factors and outcomes to be analyzed. Many of the factors for successful resettlement recur despite the cases being different in implementation period and location. The book presents highly original findings gathered by local researchers in the field directly talking to resettlers who were relocated more than a decade ago.

This original book is a unique resource for researchers and postgraduate students of development studies, environment, geography, sociology, and anthropology. It also makes policy recommendations for future resettlement programs that are of great value to development policy makers, planners, water-resources engineers, and civil society protest groups.

Ryo Fujikura is Professor at the Faculty of Humanity and Environment, Hosei University, Japan.

Mikiyasu Nakayama is Professor at the Department of International Studies, Graduate School of Frontier Sciences, the University of Tokyo, Japan.

Routledge Studies in Development, Displacement and Resettlement

Resettlement Policy in Large Development Projects

Edited by Ryo Fujikura and
Mikiyasu Nakayama

Routledge
Taylor & Francis Group
LONDON AND NEW YORK

earthscan
from Routledge

First published 2015
by Routledge
2 Park Square, Milton Park, Abingdon, Oxon OX14 4RN

and by Routledge
711 Third Avenue, New York, NY 10017

First issued in paperback 2018

Routledge is an imprint of the Taylor & Francis Group, an informa business

British Library Cataloguing-in-Publication Data
A catalogue record for this book is available from the British Library

Library of Congress Cataloging-in-Publication Data
Resettlement policy in large development projects / edited by Ryo Fujikura
and Mikiyasu Nakayama.
pages cm. — (Routledge studies in development, displacement and
resettlement)
1. Migration, Internal—Developing countries. 2. Land settlement—
Developing countries. 3. Dams—Developing countries. 4. Construction
projects—Developing countries. I. Fujikura, Ryo, 1955- II. Nakayama,
Mikiyasu.
HB2160.R47 2015
333.3'1091724—dc23
2014048172

ISBN 13: 978-1-138-59786-0 (pbk)
ISBN 13: 978-0-415-74997-8 (hbk)

Typeset in Sabon LT Std
by Swales & Willis Ltd, Exeter, Devon

'The overview and detailed insights provided in this book make it very important, timely and informative. The examples are wide ranging and the analysis critical. Professionals responsible for the technical and environmental outcomes of dam construction will find the material up to date and relevant. Those needing to know about social and livelihood impacts of large infrastructures will be especially assisted by the analysis.'

Professor Tony Allan of King's College London and the School of Oriental and African Studies, London

'Dams are an important temporary solution to the management of water resources and the generation of energy. Perhaps their highest initial cost is the cost of resettlement. This carefully researched volume examines that cost in detail. It is essential reading for policymakers involved in decisions about new dam construction.'

Michael Menaker, PhD
Commonwealth Professor
The Department of Biology
University of Virginia, USA

Contents

Figures

Tables

Boxes

List of contributors

All chapters were edited by Ryo Fujikura and Mikiyasu Nakayama with contributions from the following authors, listed alphabetically by chapter:

Chapter 1
Ryo Fujikura, Mikiyasu Nakayama, Naruhiko Takesada

Chapter 2
Miko Maekawa, Kyoko Matsumoto, Mikiyasu Nakayama, Naruhiko Takesada, Dian Sisinggih, Bounsouk Souksavath, Sunardi

Chapter 3
Syafruddin Karimi, Jagath Manatunge, Kyoko Matsumoto, Dorothea Agnes Rampisela, Bounsouk Souksavath

Chapter 4
Ti Le-Huu, Syafruddin Karimi, Miko Maekawa, Chi Cong Nguyen, Bounsouk Souksavath

Chapter 5
Erhan Akça, Ryo Fujikura, Hidemi Yoshida, Dorothea Agnes Rampisela

Chapter 6
Dorothea Agnes Rampisela, Mikiko Sugiura, Hidemi Yoshida

Chapter 7
Ryo Fujikura, Mikiyasu Nakayama

Notes on contributors

Erhan Akça is an associate professor at Adıyaman University, Turkey. He received his MSc and PhD from Çukurova University, Turkey. He has published more than 110 articles, books, and chapters on environmental and natural resources management, and is now working on archaeometry, soil science, and soil technology.

Ryo Fujikura is a professor at Hosei University, Japan. He received his BSc and MSc from the University of Tokyo, Japan, and Doctor of Natural Science degree from Innsbruck University, Austria. His current research topics include environmental policy formulation, and the environmental and social impacts of infrastructure development.

Syafruddin Karimi is a professor of economics at Andalas University, Indonesia. He received his MA (economics) from the University of the Philippines, and PhD (economics) from Florida State University, United States. His research interests include land inequality and poverty, and the social and economic impacts of natural disasters, resettlement, and infrastructure development.

Ti Le-Huu is an adjunct professor at Danang University of Technology, Vietnam. He completed his MEng and DEng at the Asian Institute of Technology, Thailand, and was formerly the chief of the Energy Security and Water Resources Section of the United Nations Economic and Social Commission for Asia and the Pacific (UNESCAP). His research includes water security of national and international river basins.

Miko Maekawa is the director of the office of regional funds at the Sasakawa Peace Foundation. She graduated from Sophia University, Japan. She holds an MSc (environment and development) from the University of East Anglia, UK, and a PhD (international studies) from the University of Tokyo, Japan.

Jagath Manatunge is a senior lecturer at the University of Moratuwa, Sri Lanka. He received his BSc (civil engineering) from the University of Moratuwa, Sri Lanka, MSc (environmental technology) from Imperial

College London, UK, and PhD (environmental sciences) from Saitama University, Japan. His research interests include water and wastewater technologies, pollution control, and environmental management.

Kyoko Matsumoto is a senior safeguard officer at the Japan International Cooperation Agency. She received her MSc from Oregon State University, United States, and Doctorate (international studies) from the University of Tokyo, Japan. Her current research interests are environmental and social impacts of infrastructure development projects and regulations on transboundary impact assessments.

Mikiyasu Nakayama received his BA, MSc, and PhD from the University of Tokyo, Japan. He has worked at the United Nations Environment Programme, Utsunomiya University, Japan, the World Bank, and Tokyo University of Agriculture and Technology. He currently serves as a professor in the Graduate School of Frontier Sciences at the University of Tokyo.

Chi Cong Nguyen is a professor and the dean of the Department of Water Resources, Danang University of Technology, Vietnam. He completed his PhD at the French Institute of Science and Technology for Transport (IFSTTAR), Nantes University, France. His research interests include water resources management and hydrology for river basins.

Dorothea Agnes Rampisela is an associate professor of Hasanuddin University, Indonesia. She received her MSc and PhD (hydrology) from Kyoto University, Japan. Her current work is in transdisciplinary research on water management, including environmental and social impacts of infrastructure development.

Dian Sisinggih is an associate professor at the Department of Water Resources Engineering, University of Brawijaya, Indonesia. He received his PhD (integrated river basin management) from the University of Yamanashi, Japan. His research interests include the water-sediment environment and community-based disaster risk management.

Bounsouk Souksavath is an associate professor at the National University of Laos. He received his BEng from the University of Lvov, Ukraine, and his MSc and PhD (international environmental studies) from the University of Tokyo, Japan. His current research interests include environmental impact assessments for infrastructure development projects in Lao People's Democratic Republic.

Mikiko Sugiura is an associate professor at Sophia University, Japan. She received her Bachelor of Law, Master's (international studies) and PhD from the University of Tokyo, and BA from Sophia University. Her current research subjects include irrigation management, water users' associations, and natural resources management.

Sunardi is an associate professor at the Department of Biology, Padjadjaran University, Indonesia. He completed his doctorate at the Department of Environmental Science and Human Engineering, Saitama University, Japan. His research interests cover water resources conservation, including the impacts of pollution on aquatic ecosystems and human health.

Naruhiko Takesada is a professor at Hosei University, Japan. He received his BA, MA, and PhD (international cooperation) from the University of Tokyo. He has 14 years' experience working in a Japanese aid agency. His research topics include development ethics, and the environmental and social impacts of development intervention, especially displacement.

Hidemi Yoshida is an associate professor at Hosei University, Japan. Her research areas include microfinance, rural development, and poverty alleviation through social entrepreneurship. She received her MA (international studies) from Saitama University, Japan.

Foreword

Thoughts while reading this new book

This monograph authored by Professors Ryo Fujikura and Mikiyasu Nakayama—two eminent Japanese anthropologists and development specialists—comes out at a time when the problems of forced population displacement and population resettlement are at the forefront of the international development agenda more than ever before. The authors invested an entire decade—from 2006 to 2015—in carrying out tenacious field research and theoretical analyses to communicate their empirical findings and policy conclusions. They are rewarded now by a most propitious moment for their book to land on universities' desks, in bookstores, and also in the hands of development managers and practitioners.

The political economy of expropriation, dispossession, and forced displacement, and their mass-scale impoverishment effects, have generated complex and dramatic problems that are accumulating now, unsolved, on the agendas of states, governments, and development agencies throughout the developing world. This book brings valuable knowledge and highlights constructive approaches that can help address these unsolved problems better than they are currently addressed.

In responding to the authors' invitation to introduce this monograph and share some thoughts about its complex topic, I will briefly comment on this book's unusual and productive methodology and attempt to place the volume in the context of the international resettlement literature and of the particular moment in time at which this original book is coming into the public light.

Fujikura and Nakayama's approach can be characterized succinctly by three defining dimensions: (1) it centrally deals with the *policies and legal normative frameworks* that must guide the governance of development-caused deliberate but compulsory population resettlement, as the book's title tells us from the outset; (2) it focuses on hydropower dam construction as one of the major sources of today's *global problem of development-caused displacement and resettlement* (henceforth DCDR); and (3) it has chosen the Asian continent as the geophysical and sociopolitical context for their anthropological, economic, and cultural research. The purpose that inspired their empirical investigation and analytical effort is crystal clear and admirable: to search for better solutions to the kind of forced displacement that deprives, impoverishes, and "socially excludes" people from their productive activities, homes, and hearth (Sen 2000).

The authors themselves explain to the readers that their book examines three main sets of issues: the *theories* that deal with their subject; the *operational practices* through which dam projects and their entailed displacements are currently implemented and managed or, on the contrary, are mismanaged and converted in unnecessary social disasters and uncounted individual tragedies; and the existential *challenges* that DCDR processes impose on the immediate livelihoods of tens of millions of people who—by no fault of their own—happen to be in the way of infrastructural developments.

These thematic choices by the book's authors were made 10 years ago, in 2006. Now, fast-forward to 2015: and at this point in time we realize immediately that this volume examines precisely key issues of a global simmering debate that now has exploded, is not abating but, on the contrary, keeps snowballing from one day to another, literally as this "foreword" is being written. Let us therefore recall some of the landmarks that configure the international context and the controversial directions of this unsolved debate.

Policy advances in Asia and surprise policy setbacks at the World Bank

The history of policies for population resettlement introduced in international development is recent and rather short. This history started in 1980 with the first ever such policy on resettlement, adopted by the World Bank explicitly for protecting the populations affected by the negative effects of the compulsory displacement caused by some of its financed projects (World Bank 1980). No other development agency was guided at that time by such a normative policy framework. Forced displacement operations were simply left out of development projects, which financed only the "cause" of displacement (the newly built infrastructure) but did neither assume responsibility nor provided the financing for the sound resettlement of the displaced people and the reconstruction of their livelihoods. Such displacements were mis-defined as "collateral damage" and belittled as acceptable "secondary side effects."

The World Bank's 1980 pioneering policy instantaneously became, by its content, a "game changer" in development practice: it incorporated the resettlement of the displaced people as an *integral* and *mandatory* component of the infrastructure project that caused displacement. Moreover, to be truly effective in actual practice the policy predicated the appraisal and approval of such projects on the prior preparation by the borrowing country of a plan for relocating the affected population, to be incorporated in the project itself.

This pioneering change, however, was not immediately replicated by other development agencies. Even less, it was considered only reluctantly by the borrowing countries that did not have—and declined to adopt—comparable domestic resettlement policies.

Implicitly, that pioneering policy opened an explicit controversial area and era in international development: the era of intensifying struggles around two major imperatives—the *internationalization* of comparable policies in the all the world's major development agencies, and the *internalization* of such policies within the domestic legal frameworks of the developing countries themselves.

During the following three decades (1980–2010) the "two imperatives" evolved un-evenly: the "internationalization process" made enormous progress through a "ripple effect" during which the World Bank policy served as a template embraced increasingly toward virtual generalization at the international level; conversely, the "internalizing process" at the domestic level has advanced relatively little and insufficiently (Cernea 2005).

At the beginning of the current decade (that is, when the present book's field research was still ongoing), the vast majority of developing Asian countries (including Indonesia, Sri Lanka, India, the Mekong countries, Bangladesh, Pakistan, and others) were still lacking such domestic resettlement policies (there was only one major positive exception: China, which since the mid-1980s began to craft and build up its domestic architecture of rural and urban legislation on development-caused resettlement).

Asia's dysfunctional policy vacuum on DCDR prompted Fujikura and Nakayama to focus this present book on resettlement policy issues and to substantiate through empirical research and conceptual argument the need for such policies in Asia. The numerous disastrous displacements, the obvious impoverishment of the disenfranchised people deprived of legal protections, and particularly the physical and political growing resistance of the affected populations have begun to wrestle out changes from some reluctant governments of some Asian countries. And, indeed, the present book's argument is itself being validated by recent significant changes in some domestic policies.

The most remarkable and historic such policy change is India's recent adoption (2013) of its first ever national "Resettlement and Rehabilitation" Policy Act. After public/political debates that lasted for many decades, the country's Parliament adopted the first law regulating how to carry out the population resettlement caused by eminent domain land acquisition. Until then, the law that provided a legal umbrella to mass-scale displacement—while legalizing injustice—was the obsolete colonial Land Acquisition Act (LAA), dating from 1894, long criticized for serial violation of human rights and deeply inequitable economic effects.

A new law became necessary and, at long last, the LAA was replaced in October 2013 with a new act titled like a manifesto: The Right to Fair Compensation and Transparency in Land Acquisition, Rehabilitation and Resettlement Act (LARR—see India 2013). The new LARR pivoted on four novel and crucial elements: some degree of consent; upfront social impact assessment; improved compensation; and some rehabilitation provisions. Despite this progress, before the implementation of the new law could begin in earnest, the government newly elected in India in 2014 began modifying the LARR, stripping it of several essential provisions that introduced some degree of consent and were apt to identify upfront the risks and adverse impacts that would impoverish the displaced people (*The Hindu* 2014; Iyer 2015; Guha 2015).

Other policy developments in Asia also exemplify the new trend: in Sri Lanka, a policy document was adopted, though it did not go far enough in providing sound protections; in Indonesia, a reexamination is now ongoing on how the eminent domain principle can be used; and in China, a new significant policy decision was adopted in November 2012. The State Council of China decided that China's new industrial projects must carry out a preliminary identification and analysis of the "risks to social stability" that such industrial projects might cause. China's Minister for the Environment stated: "No major projects can be launched any longer without social stability risks evaluations. By doing so, I hope we can reduce the number of mass incidents in the future" (Bradsher 2012). The specificity of this type of risk analysis speaks eloquently about the political risks of underestimating the economic harm that displacement inflicts (Cernea 2015b).

Despite such significant policy steps at the level of some countries, however, there are also contrary trends at the international level. A major pushback has come from an unanticipated corner: the World Bank. This erstwhile pioneering agency in resettlement has engaged, starting in 2012, in a "revision and update" of its resettlement policy and other safeguard policies. Yet, to virtually worldwide surprise, the first phase (2014) of this exercise of "updating" turned into an exercise of dilution and downgrading, hollowing out much of the essential content of the resettlement policy.

The response has been a loud and broad outcry against this unjustified twist. It has picked up steam on a worldwide scale among civil society organizations (Oxfam 2015), scholarly associations (Cernea 2015a), environmental institutions, and a large plethora of many other actors who advocate strengthening of the resettlement policy, and reject any dilution of it.

This exercise, however, is far from being completed, and its final results are still difficult to anticipate at this time. The opposition to such dilution keeps growing internationally, as the present book goes to print. Fujikura and Nakayama's book was completed before this policy downgrading came to be known publicly but the book's fundamental argument clearly weighs in in favor of strengthening, not diluting, the content of resettlement safeguard policies. And it is fair to say that the relevance of Fujikura and Nakayama's book to the ongoing international debate is its richly documented demonstration that sound social policies on resettlement are indispensable to avoid the impoverishment of the people adversely affected by displacement.

Japan's experience and innovations toward socially sound resettlement

Essential for the major contribution of the present book to the international resettlement literature is, in my view, the richly textured empirical documentation of several innovative approaches tested and developed in Japan for addressing differently the central issues of land expropriation.

Not long ago, other Japanese social scientists had wistfully noted that "the sociology of development in Japan is an under-explored, if not under-valued, field of study" and that that "has hindered the growth of the sociology of development" in Japan (Hamamoto and Sato 2012). This has certainly been true; and to some extent Japan's creative approaches may still be under-explored.

However, to their credit, and to the benefit of the Japanese and international resettlement research and literature, Nakayama and Fujikura have made big strides toward bringing to public awareness some little-known Japanese approaches that are sound, practicable, and may be transferrable to other countries' circumstances. Almost one-third of the number of hydropower dam projects studied in the present book—5 out of 17—are from Japan, and the others are from developing countries; this brings into the knowledge circuit a set of practices that contain gold nuggets of creative, precious original approaches. There are few cases in the recent literature in which the authors analyze in the same book resettlement processes from both developed and developing countries.

Particularly worthy of readers' attention, in my view, is the book's description of Japan's attempts to avoid land expropriation and replace it with long-term land leases that can ensure a continuous stream of benefits to the land owners and their descendants over the duration of the dam-reservoir's life. A rent scheme was adopted for a cascade of three small-scale dams constructed in the early 1950s on the Jintsu River (Jintsu River Dam No. 1, No. 2, and No. 3). The affected families received a "one-time payment" when the land was leased to the hydropower company; in addition, new houses for the resettlers were constructed free of charge. The one-time payment helped the families to shift to immediately create an alternative production or income source for themselves, while the rent for lands leased added annually to an amount that enabled the landowners to share in the project's stream of benefits. Rents have been paid for more than 50 years after project construction, preventing the de-capitalization and impoverishment of the landowners. When Fujikura and Nakayama carried out their surveys, they interviewed in many cases the second generation of the land-owning families and in some cases even the third generation. That confirmed that the approach enabled the former farming population to overcome the immediate practical loss of land and to improve and rebuild their livelihoods compared their pre-project level.

Equally interesting is the discussion in the present monograph of the resettlement planning for the Namata Hydropower Dam—a dam that, eventually, was not even built. That high dam, planned for the Tone River basin in the 1950s/1960s, was to displace over 3,000 families, totaling some 10,000 people. To ensure resettlement success, very detailed planning was done long before dam construction had to start. The guiding idea was to prepare a strategy for reconstruction and development even *before* the affected persons were exposed to the loss of assets, so that at the time of relocation, they could already "hit their new ground running" avoiding a long recovery time and instead stepping directly into a prepared developing context. For extraneous reasons, the Namata Dam was eventually not built; however the experiences of its planning remain a usable example today of how resettlement could be planned early, even before the displacement starts, as a reconstruction-with-development process. This approach comes close to what today in China is defined as "resettlement with development" instead of "resettlement with initial impoverishment."

Shifting the research focus toward reconstruction

Fujikura and Nakayama rely in their book not only on their own research, but also abundantly use in their reasoning existing theories, models, concepts, and relevant findings from empirical studies authored by other researchers from different countries. Yet they also critique the general resettlement literature for "insufficient attention to medium and long-term processes of post-displacement reconstruction." I strongly support this critique (Cernea 2012).

Much of the contemporary research literature limits itself narrowly to a short timespan: it describes the displacement segment's problems and covers only a limited interval after relocation. With this compressed exposure and timeframe, most studies are prevented *ab ovo* from analyzing the slow and lengthy processes of recovery after displacement, which are prone to patterns of typical and idiosyncratic risks, difficulties, painful failures, and hard-to-win successes. This way, research remains behind what must be studied for understanding better the anatomy and potential engines of post-displacement reconstruction.

Considering the current state of resettlement scholarship and findings, I think the most powerful message that empirical evidence communicates is that the outcome of displacement based on eminent domain is, predominantly, the immediate impoverishment of those displaced. This happens not just in a few developing countries but virtually all over the developing world. Impoverishment as process and newly produced "poverty groups" are undeniable realities. The current economics of displacement and reset-tlement is deeply flawed and needs to be structurally changed through investments and benefits-sharing (Cernea 2008). This skewed, unjust, and immoral economics externalizes a part of the projects' costs upon the people who are deprived by fiat of their life-sustaining assets. Primarily the rural poor, whose land ownership, production systems, and income sources are demolished, become instantaneously and long impoverished, economically and socially excluded from development.

In consensus with this book's authors, I think that research on displace-ment must be complemented by increased, extensive, and in-depth research on *people's post-displacement existence* and up-hill struggles. Research must not only continue to document the risks and realities of displacement that breeds impoverishment and human tragedies, but also it must move fur-ther. It must analyze and document peoples' post-displacement-lives, reveal recovery's enormous difficulties, its risks patterns, the resources for and the barriers to effective reconstruction, to reinforce through such research the argument for changing today's predatory economics of involuntary resettlement.

To overcome the tunnel-vision trained on displacement alone, and con-sistent with their methodology, Fujikura and Nakayama decided to select for their surveys mostly hydropower dams completed between 20 and 50 years before they started their surveys. They devoted primary attention to examine what enables displaced populations *to overcome* the losses and hardships of displacement and to rebuild their livelihoods. Some of the individuals they interviewed still belong to the population that was dis-placed, but in many cases the second generation is now the main income earner in households; in some dams, it is the third generation. The authors willingly traded the traditional "participant observation" on displacement for the cognitive benefits of a longer time perspective over the after-displacement years. This also enabled them to go beyond the economic variables; they devote a full chapter to the human emotional and cultural effects of displacement. Some intangibles that are hard to quantify but that indelibly color the daily existence of those affected even in the second and third generations are captured insightfully in this monograph's perceptive analyses. Overall, the book successfully embodies authors' goal to offer a "post-project evaluation of the long term effects" of development-caused displacement.

What, in my view, gives added value to this *ex-post* research is that the authors pursued not only the rendition of significant empirical facts and of characteristics particular to one or another country or process but also they have strived to distill generalizations apt to enrich the conceptual and theo-retical analysis of these complex processes. The diversity of the empirical situations encountered has enabled the authors to make analytical compari-sons and to assess the differential values between various approaches to common and fundamental issues in resettlement policies and practices.

Will development-caused displacement decrease or increase?

The correlation between the *frequency* of displacement processes and the number of projects investing in infrastructure is direct. Similarly direct is the correlation between the *total numbers* of people displaced and the total amount of financial investments in infrastructure. In short, the higher the number of projects, and the larger the investments, the more significant DCDR becomes in today's and tomorrow's world.

Macroeconomic studies and forecasts anticipate that the world at large is moving toward "the biggest investment boom in human history" (Flyvbjerg 2014). The total global megaproject spending is anticipated at US $6 to US$ 9 trillion annually, representing 8 percent of the total global gross domestic product. This will be achieved through a continuous increase in the size and frequency of megaprojects on the worldwide scale (Flyvbjerg 2014).

Encrypted in these macro-forecasts is the response to the question asked above: Will development-caused displacement decrease or increase? The response is obvious. And sobering.

Michael M. Cernea[1]
April 18, 2015
Bethesda, MD, USA

Note

1 Michael M. Cernea (USA), sociologist, received his PhD from the University of Bucharest, and in 1970–1971 was a fellow at CASBS. In 1973 he was elected Vice-President of European Society for Rural Sociology, and joined the World Bank (1974) as its first in-house sociologist. Cernea was the proponent and author of World Bank's first-ever Policy on Involuntary Resettlement (1980), broadly replicated internationally. He is a member of Romania's Academy of Sciences and currently works as a senior fellow at Brookings Institution, Washington DC.

References

Bradsher, Keith (2012). "Social risk" test ordered by China for big projects, *The New York Times*, 12 November. Retrieved from: www.nytimes.com/2012/11/13/world/asia/china-mandates-social-risk-reviews-for-big-projects.html?_r=0.

Cernea, Michael M. (2005). The "ripple effect" in social policy and its political content: a debate on social standards in public and private development projects. In Michael B. Likosky (ed.) *Privatising Development: Transnational Law, Infrastructure, and Human Rights*, Leiden: M. Nijhoff, pp. 65–101.

Cernea, Michael M. (2008). Compensation and investment in resettlement: theory, practice, pitfalls, and needed policy reform. In Michael M. Cernea and H. M. Mathur (eds.) *Can Compensation Prevent Impoverishment?* Delhi: Oxford University Press, pp. 15–98.

Cernea, Michael M. (2012). *The Reconstruction Challenge in Development-Caused Displacement*. Keynote Address, International Conference of Displacement and Resettlement, Oxford University, UK.

Cernea, Michael M. (2015a). *The Economic and Moral Imperatives for a New Policy Vision and Stronger Safeguards.* Keynote Address, Opening Plenary of the INDR Sessions on Social Safeguard Policies, SfAA, Pittsburgh, March 24–28.

Cernea, Michael M. (2015b). Landmarks in development: the introduction of social analysis. In Susanna Price and K. Robinson (eds.) *Making a Difference? Social Assessment Policy and Praxis and its Emergence in China*, New York: Berghahn, pp. 35–59.

Flyvbjerg, Bent (2014). What you should know about megaprojects and why: an overview, *Project Management Journal*, 45 (2), 6–19.

Guha, Abhijit (2015). Dangers of Indian reform of the colonial land acquisition law, *Global Journal of Human Social Science*, 15 (1), 1–8.

Hamamoto, Atsushi and Yutaka Sato (2012). Contemplating the sociology of development: genealogy and approaches in Japan and the West, *Journal of International Development Studies*, 21 (1/2), 11–29.

India (2013). The right to fair compensation and transparency in the Land Acquisition, Rehabilitation and Resettlement Act, 2013, *The Gazette of India*, September 27.

Iyer, Ramaswamy R. (2015). When amendment amounts to nullification, *The Hindu*, January 15.

Oxfam (April 2015). The suffering of others: the human cost of the International Finance Corporation's lending through financial intermediaries, Oxfam Issue Briefing. Retrieved from: http://reliefweb.int/sites/reliefweb.int/files/resources/ib-suffering-of-others-international-finance-corporation-020415-en.pdf.

Sen, Amartya (2000). Social exclusion: concept, application, and scrutiny, *Social Development Papers No.1*, Manila: Office of Environment and Social Development, Asian Development Bank.

The Hindu (2014). Nod for ordinance to amend Land Act, December 29. Retrieved from: www.thehindu.com/news/national/cabinet-approvesordinanceto-amend-land-acquisition-act/article6735783.ece.

World Bank (February 1980). *Social Issues in Involuntary Resettlement under Bank-Financed Projects*. OMS 2.33, Washington, DC: The World Bank.

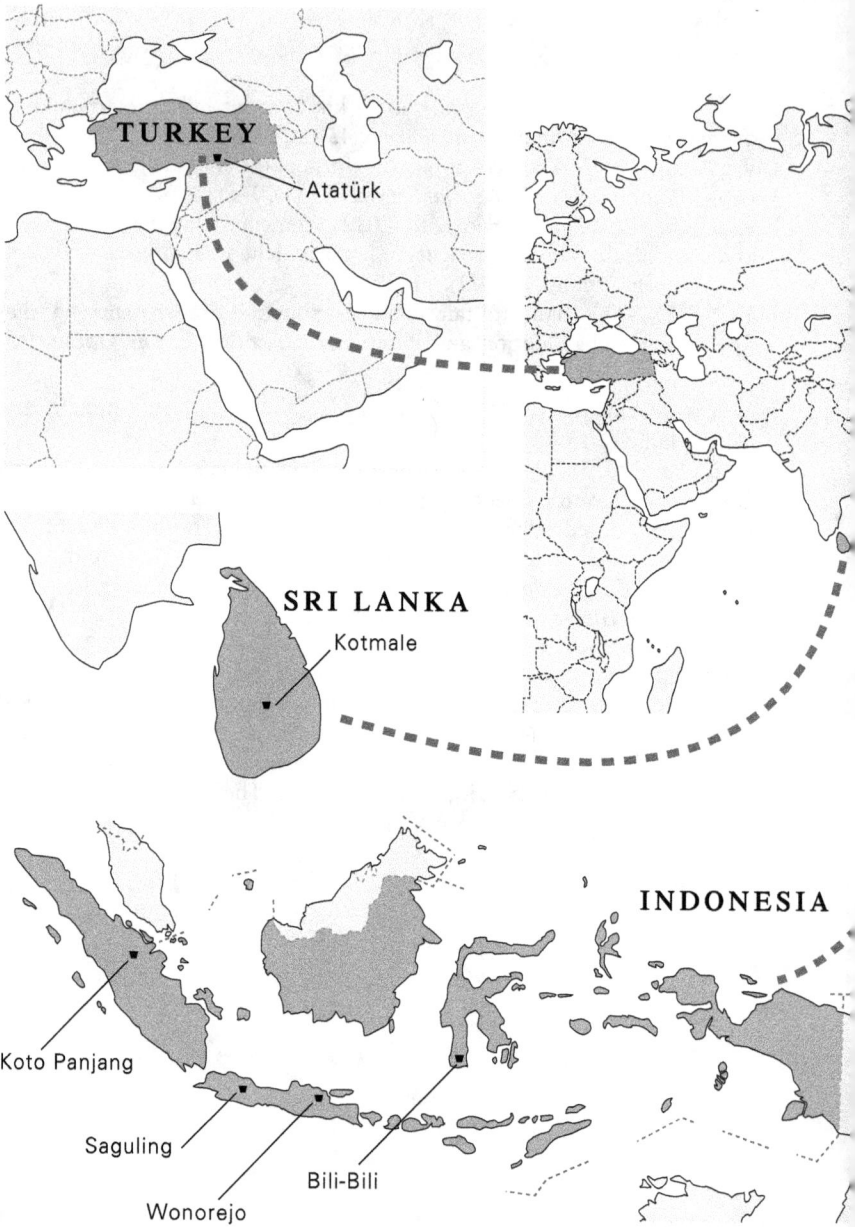

Figure 0.1 Locations of dams and resettlement case studies.

settlement case studies

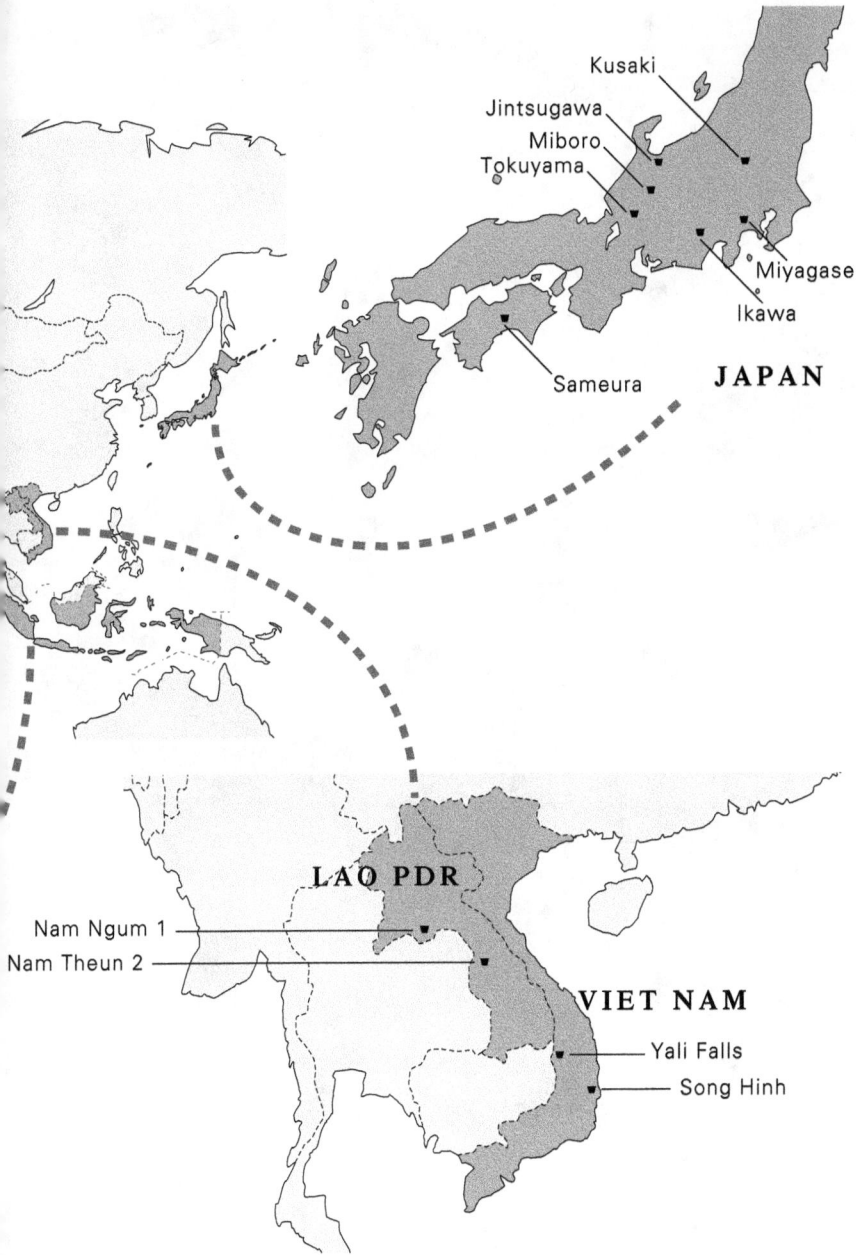

Kusaki

Jintsugawa
Miboro
Tokuyama

Miyagase

Ikawa

Sameura

JAPAN

LAO PDR

Nam Ngum 1
Nam Theun 2

VIET NAM

Yali Falls
Song Hinh

1   Introduction

Issues and brief history of dam-induced resettlement

The idea of large dam development appears to be nearly obsolete in developed countries where people generally enjoy an adequate supply of water and electricity. In the developing world, however, many countries plan to build dams, seeking a solution to insufficient and unstable water supplies—for domestic, agricultural, or industrial use. As well, most developing countries see dam development for hydroelectric power generation as a feasible option for clean energy development. While dams are generally perceived in developed countries as relics of the modernization process, developing-country governments are generally unwilling to discard the option. Dam developments, however, inevitably have adverse and irreversible impacts on the natural and human environments. These impacts are a major driver of opposition to dam construction in most developed countries. Where there is less public scrutiny and depending on their location, dams can displace people and communities when land is submerged by reservoirs. Dam construction can have impacts on communities and the social environment.

Dam-induced resettlement often changes the lives of resettlers dramatically. People often have to leave their ancestral land and homes, find new jobs in a new place, and become part of an unfamiliar or newly-built community. Even with reasonable compensation, one could imagine that it would take many years for these people to restore their livelihoods and enjoy the stability they had before resettling.

Dam-induced displacement has occurred in both developed and developing countries. For example, in the United States, the Norris Dam, completed in 1936 as the first dam under the Tennessee Valley Authority, displaced more than 3,500 households (McDonald and Muldowny 1982, p. 70). The Sup'ung Dam, built between 1937 and 1944 on the border of Manchuria and Korea (then a colony of Japan) under the auspices of the Japanese colonial government, induced resettlement of approximately 15,000 households and 70,000 people on both sides of the river (Hirose 2003). In Ghana, the Akosombo Dam displaced about 80,000 people, as the new country demanded electricity and industrial development for economic independence. In many developing countries that obtained political independence after World War II, dam construction was a symbol of modernization, and dams were often depicted on monetary notes as a display of national pride. In Japan as well, during the country's post-war period of rapid economic growth, water and electricity supply was supported by the construction of a number of dams around the country. In 1951, hydropower provided 79.8 percent of Japan's electricity supply, but that supply came at the cost of many cases of dam-induced displacement.

The number of people displaced and resettled by dam construction world-wide has been significant. In 1994, an analysis by the World Bank found that at the time, more than four million people were being displaced every year by the construction of over 300 dams worldwide. The cumulative toll of displacement from dams, urban development, and transportation development in the preceding ten years was estimated at 80 to 90 million people (World Bank 1994). A report by the World Commission on Dams (WCD) estimated that the all-time cumulative displacement caused by dam construction worldwide was 40 million to 80 million people (WCD 2000, p. 104).

Two points must be noted on the extent of dam displacement. First, exact numbers of displaced persons are not available (Scudder 2005, pp. 21–22). This fact suggests that society as a whole has not paid adequate attention to those who have been displaced under the banner of development, or has even intentionally avoided studying the impacts. In other words, the global community has not been caring enough toward individuals.

Second, the magnitude or scale of displacement tends to be taken mainly as an administrative consideration. It is true that in developing countries, the large scale of displacement makes issues more complex. For example, the larger the displacement, the more difficulty there is in finding alternative land for displaced people to reside on and cultivate. Besides that, however, the displacement could have huge negative impacts on each individual's life, irrespective of the total number of resettlers. One should note this point in both research and practice. This is one of the reasons why this book addresses psychological and subjective issues relating to resettlement.

The general view toward dam-induced displacement experienced a great turning point in the 1980s. Since then, related issues have attracted major attention, not only locally but also in international society. Previously, dam-induced displacement was perceived as a side effect or necessary sacrifice in the course of development or modernization. Then Indian Prime Minister Nehru's statement, "If you are to suffer, you should suffer in the interest of the country" (Viegas 1992, p. 53), reveals the typical narrative of those times. However, by the 1980s it was well known that displacement had adversely impacted resettlers, especially in developing countries. Actions aiming for development had created poverty for some. This outcome was not only ironic, but also tragic. People began criticizing dam construction that involved resettlement.

The controversy over development plans in India's Narmada Valley in the 1980s was one of the turning points. The government of India had conceived of a plan for water-resources development there as far back as the 1950s. In 1985, one of the dam construction plans was initiated, with support promised by the World Bank and bilateral donors, including the Japanese government. The dam was expected to benefit more than a million people by supplying water, but with more than 100,000 people to be displaced.[1] Demonstrations and protests escalated in India. Through international networks and the media, the issue was picked up and drew the attention of civil society. International non-governmental organizations (NGOs) put rigorous pressure on the World Bank and bilateral donors. Finally, when the World Bank requested that very substantial changes and measures be implemented to improve the displacement/resettlement process, the Indian government withdrew its loan requests for the dam construction.

This case can be seen as a turning point in two ways. One is that the magnitude of displacement became the focus of attention. Since this case, dam-induced displacement has become a major issue in water-resource developments in many other places. The other is that donors' intervention in displacement and resettlement was justified and strengthened. Previously, the issue of displacement or land acquisition had been perceived purely as an internal responsibility of developing countries. For this reason, donor organizations had not participated in supporting sound resettlement, even if they financed the project that was causing displacement. Before Narmada, only the World Bank had adopted a formal policy for guiding involuntary resettlement operations. After the Narmada case, in the early' 90s, however, all other donor agencies began to adopt policy guidelines and procedures modeled after the World Bank policy for displacement and resettlement; they also started to demand developing-country governments to prepare laws and regulations regarding land acquisition and compensation. A project causing displacement would then be carefully scrutinized in the planning process and the government would have to prepare a detailed compensation and resettlement plan so as to receive funding for projects.

Once the focus of attention turned to dam-induced displacement, did changes result? When both developing-country governments and donors are well equipped with regulations and procedures, does displacement result in less impoverishment? We know now that, unfortunately, the answers to both questions are "no." At a glance, a change in awareness after the Narmada controversy did result in a better environment for those who are displaced by dam construction. Especially, there seems to be more communication between dam proponents and opponents in order to avoid further hardships for resettlers. A desirable cycle of attention–communication–improvement appeared to become a part of the development process. It did not work in the expected way, however. While dam-development proposals continued to appear, criticism and opposition were constantly present. Better procedures may be in place, but people are still being displaced as before. In each case, the lives of displaced persons are put in turmoil. It was felt that more meaningful communication and solutions to dam-induced displacement were necessary.

The WCD was inaugurated in 1997 with the expectation of playing a decisive role over the issue of water-resources development—not only for dam-induced displacement but also for other aspects of development. After extensive research and consultations, the WCD published its final report in

2000 as "a new framework for decision-making." Clearly recognizing the risk of displacement and rights of the resettlers, it puts forward 5 core values, 7 strategic priorities, and 26 guidelines for good practice (WCD 2000, pp. 197–307). In regard to dam-induced displacement, the report proposes the implementation of the Mitigation, Resettlement and Development Action Plan based on an adequate impoverishment risk analysis. Also, project benefit-sharing mechanisms are recommended in order for adversely affected people to secure entitlement for a share of project benefits (WCD 2000, pp. 296–301). The activity of the WCD ended with the publication of its final report, but the United Nations Environment Programme conducted the Dams and Development Project as a follow-up program for the purpose of dissemination of the recommendations and accumulation of good practice.

At this point, we raise further questions: With the WCD guidelines, are we now in a better position to solve the problems of dam-induced displacement? Are resettlers better-off if relocated in more recent dam-development projects? The answers to these questions may not be as positive as one would hope. Not only were the WCD guidelines not widely adopted, they were also clearly rejected by the national governments of some developing countries, including India. Thayer Scudder, one of the WCD committee members, criticizes the reluctance of leading agencies such as the World Bank and the World Water Council to accept the recommendations (Scudder 2005, pp. 10–13). In parallel, there have also been, and are severe criticisms of, the WCD activities and its report. Some critics targeted the WCD recommendations and guidelines as impracticable. Fujikura and Nakayama (2002) found that out of 26 guidelines, only 6 may be applicable to current practice and others are too theoretical or incomplete for operationalization in real dam-development projects. Asit Biswas (2004) criticizes the legitimacy of the WCD itself in regard to the selection procedure of committee members. He concludes that even without the WCD the situation would not be much different (Biswas 2004, pp. 10–12).

On the other hand, there have been positive comments about the WCD as well. Goulet (2005) makes one such statement in his study. He answers criticism of the WCD by saying that the participatory process advocated in the WCD report did have a meaningful influence on governmental policy, beyond individual projects, by illustrating a Brazilian case where stakeholder participation had influenced national water policy. Bradlow (2001) also points out that the WCD showed that for dealing with decision-making in the context of development, the potential for success lies only in the "modern view," which involves participatory processes and is different from the "traditional view" managed solely by experts and bureaucrats.

Regarding institutional issues, Cernea et al. (2008) conducted an institutional assessment of resettlement at the Ilisu Dam, Turkey, and emphasized the crucial role of implementing agencies and the importance of coordination among related agencies. They reveal that poor coordination among eight ministries and agencies designated to collaborate in the implementation of the resettlement resulted in a lack of information and awareness.

Using a metaphor, "Many fingers, but not of the same hand," they conclude that highly compartmentalized and weakly integrated institutions hindered resettlement implementation. While considerable amounts of human and budgetary resources were devoted to the technical and structural components of the dam project, resources devoted to population resettlement components were far less than required. The authors argue that resettlement on such a large human scale should not be implemented simply as a "project component," which inevitably becomes constrained and under-treated as a "project within a project" but needs to be carried out as a distinct development project in its own right.

Having briefly reviewed the history of dam-induced displacement, one may come to the view that even today, dam-induced displacement continues to happen and that it is unwise to stop exploring better compensation and resettlement schemes to help resettlers avoid losses and disadvantages. This context and these perceptions are the basis of our research program, which searches for better measures for the livelihood reconstruction of displaced people.

Theory, practice, and challenges

A surprisingly large volume of research and literature deals with displacement caused by dams or other projects. In this section, we briefly review that material in the light of impacts on the practice of dam-induced displacement.

Research and literature on development-induced displacement falls into two streams (Dwivedi 2002). Dwivedi calls one a "reformist-managerial view" and the other a "radical-movementist view." The former sees development (and/or development projects) as being necessary or inevitable, even if they require the displacement of people, and tries to minimize the adverse effects of displacement by improving the process and contents of compensation and rehabilitation for resettlers. The latter sees development (and/or development projects) with displacement as the ugly face of development, causing an unfair distribution of costs and benefit, and criticizes not only displacement but also development itself. The former is often the view taken by governments and aid agencies, while the latter is often the view of NGOs that protest against development projects. As there is a gradation of claims and ideas between these two poles, the situation becomes more nuanced. Nevertheless, it seems that between these two viewpoints exists a gulf that cannot be easily bridged (Dwivedi 2002).

Empirical studies usually focus on the question of which process or content should be improved if better consequences of resettlement are to be achieved. Some studies pay further attention to what resulted in better consequences where successful cases are perceived. Both often have a reflective framework so as to improve praxis.

In this book, the focus is mainly on research on the "reformist-managerial view," since the ground reality in development activities makes necessary considerable improvement in the process and contents of compensation and livelihood reconstruction for those who are displaced. Although it is acknowledged that the literature in the "radical-movementist view" gives us opportunities to reflect on development enterprises and different perspectives on the issues, it may be asked if the view has been strong enough to achieve its goal, namely to change the course and philosophy of development. Even with persistent efforts, the view does not seem to have achieved substantial change in development practices. For example, in the Narmada case mentioned above, protests gained victory in the sense that the planned international and bilateral assistance was withdrawn and better procedures and regulations were introduced in aid agencies but that victory did not stop the Indian government from building the dam and relocating people.[2] Dams and development projects involving displacement are still being implemented today, even if they are challenged by opponents.

Two theoretical frameworks conceptualize dam-induced displacement and lead us to consider existing challenges in displacement and resettlement practice in developing countries. The first one is the Four-Stage Framework (hereafter referred to as FSF) by Thayer Scudder. The framework "theorized on how the majority of resettlers can be expected to behave during a successful resettlement process" (Scudder 2005, p. 31). Later, stages were renamed as Planning and Recruitment (first), Adjustment and Coping (second), Community Formation and Economic Development (third), and Handing Over and Incorporation (fourth) (Scudder 2005, pp. 29–44). He explained that his theoretical framework deals with successful resettlement processes rather than failed processes as he replied to criticism of his theory. It should be noted that the transition from the second stage to the third stage is the most important and interesting part. Scudder sets out two necessary conditions for this transition. One is the resettlers' radical behavioral change. The other is development opportunities (Scudder 2005, p. 37). The FSF sets its focus on human behavior during the stages of the resettlement process for up to two generations. The FSF also has its weaknesses as a theoretical framework, since it is not sensitive enough to cater for the diversity of resettlers' choice and lives, as Scudder himself admitted (2005, p. 41).

The "Impoverishment Risks and Reconstruction Model" (hereinafter referred to as IRR model) is a conceptual and operational framework and complementary tool for planners, guiding them to identify and address the pauperization risks of the displacement process. It was developed by Michael Cernea, who led the World Bank's social development work, based on worldwide empirical evidence including and also beyond field experience of World Bank financed projects. It is seen by some not only as a theoretical framework but also as a practical tool for planning compensation and post-displacement reconstruction. It has enabled other aid agencies and developing countries to have a useful tool for planning and implementing systematic as well as standardized compensation and resettlement programs.

Key concepts in the IRR Model are *risk*, *impoverishment*, and *reconstruction*. Reviewing these concepts shows inherent difficulties in the IRR Model (or inherent difficulties in displacement and resettlement). First, while it is undeniable that the IRR Model's "predictive function" (Cernea 2000, p. 21) is quite useful to formulate policy and programs regarding compensation and resettlement, its usefulness is limited as the risk assessment is done only before displacement. There are always unpredictable events in the course of reconstruction. This uncertainty is not considered adequately if one exclusively relies on the IRR Model to prepare compensation and resettlement plans.

Second, *reconstruction* tends to be reduced to recovery of income. However, while physical relocation or displacement changes broad aspects surrounding resettled populations including both natural and social conditions, evaluating livelihood reconstruction tends to focus on the financial terms, namely income restoration. Cernea stresses in recent years that compensation alone has not been enough to reconstruct resettlers' new productive basis and that benefit-sharing and/or additional investments are necessary for smooth livelihood reconstruction (Cernea 2003). In practice, reconstruction is still mainly a financial term used to measure the success of livelihood reconstruction of resettlers.

Last, confidence in economic rationality also requires attention. In the IRR Model, risk reversal is an important part in the operationalization of the resettlement action plan (Cernea 2000, p. 20). For example, to reverse the impoverishment risk of landlessness, the compensation and resettlement program is supposed to include land-based resettlement. Risk assessment is beneficial if the predictive function of the IRR Model is used well. To counteract and reverse the risks identified by the IRR model, projects need also to specify and explain these risks to the resettlers. The IRR Model itself does not guarantee that rational choice by resettlers will prevail. One can only *assume* that resettlers make rational choices when they know the risks. Since risk perception by resettlers will be different from that of planners, consultations between project sponsors and resettlers about the likely risks must take place in each project.

Scudder claims that policy planners are able to prepare and implement effective compensation and resettlement programs. He also points out that "unexpected events" have significant impacts in a number of cases (Scudder 2005, pp. 48, 70). These claims by Scudder may be well appreciated as they are based on empirical evidence through a number of case studies. However, we feel that some important issues are still left undeveloped in Scudder's elaboration. One is the problem of unexpected events. His analysis does not tell what kind of countermeasures need to be taken in the case of unexpected events. Another issue is complexity. He mentions that "a successful outcome is possible provided the necessary inputs and opportunities are there" (Scudder 2005, p. 50). But in Scudder's model, it is not clear what kind of risks appear in the resettlement and what kind of inputs or opportunities need to be provided.

Theories of displacement and resettlement for dam development have inherent constraints, especially in that those theories tend to have the purpose of serving for a better planning of compensation and the design and preparation of resettlement programs. However, in this book we propose to shift the focus from the preparation of displacement stage to reconstruction after reconstruction stage.

The theories on the resettlement are often sophisticated, making their operationalization not easy, yet necessary. The theories and the practices that we examine in this book bring up two issues in the reconstruction of resettlers' livelihoods. These are described next.

Issue 1: Tendency to see resettlers' reality in a limited perspective (inability to take diversity into consideration)

Compensation and resettlement programs based on existing theoretical frameworks tend to be based on economic rationality—usually as viewed by planners.[4] But people and their choices are always more diverse than expected, as our research program shows. Current practice may not respond well to this diversity among resettlers in consideration of administrative costs. If an evaluation result shows that income restoration has not happened or accelerated as had been expected, alternatives for further income restoration may be provided to stimulate economic rationality of resettlers. The tendency to focus on resettlers' economic rationality is strengthened if standardized and common procedures for compensation and resettlement are explored based on the IRR Model as a planning tool. This tendency seems to be an exact illustration of what Quarles van Ufford et al. described. They analyze current development policy in general from the ethical perspective, saying: "The confidence in rational design and engineering has never been greater, and the policy concepts applied reflect a growing sophistication of management which is able to absorb and deflect challenges" (Quarles van Ufford et al. 2003, p. 6).

By the same token, Dwivedi (1999) pointed out that in real life, risks are defined differently by planners compared to resettlers. He analyzed risk perception of people involved in the Narmada Valley development in India, and found that different risk perception brought about different choices by different groups of people, resulting in them either accepting compensation and displacement or opposing dam development. An important point here is that the risk acknowledged by people tends to be different from the risk perceived by planners. Usually, risk perceived by the government authority or planners is defined more objectively. The use of the IRR Model may reinforce this tendency, with the result that affected peoples' risk perception may not be taken into proper consideration.

Not only in the practical preparation stage but also in the general perception of resettlers, there is the same tendency to take an abstract view of their diversity. Nevertheless, there has been progress toward better recognizing the plight of resettlers. Improved safeguard policies and lessons derived from evaluation brought about changes consisting in the more straightforward recognition of resettlers' risks and great obstacles to recovery. In the past they were expected to make sacrifices, and the side effects of development were generally ignored. Then resettlers became the target of safeguard policies for appropriate compensation for losses and livelihood reconstruction. In recent years, an increasingly strong narrative has advocated that resettlers should be beneficiaries of the projects that displace them. In other words, there has been a shift toward talking about planning resettlement as a vehicle of new opportunities for resettlers (Mcdowell 1996, p. 7; Picciotto et al. 2001, p. 137; World Bank 2004, p. xxviii). In addition, the final report of the WCD and the literature by researchers such as Michael Cernea have emphasized the importance of benefit-sharing for displaced populations.

However, this "internalization" of displaced populations as project beneficiaries should not result in treating them as if all resettlers have the same common economic rationality. For example, the World Bank has led practices in income restoration during livelihood reconstruction for displaced populations. It is no doubt important to restore income for them, but if one's livelihood reconstruction is evaluated mainly (or only) in terms of income restoration (perhaps due to administrative feasibility), the reality as perceived by displaced persons may not be properly captured. If a person's (average) income as a factory laborer is improved after resettlement, that income restoration may be evaluated as successful or satisfactory. However, what if that person had desired to (continue to) be a farmer? Does becoming a factory worker really mean successful livelihood reconstruction? Statistically, this resettler may be counted as a successful case, but in terms of that person's meaning of life it may not be. Each person and/or family may have different constraints and strategies and, hence, different choices, and internalization of these persons as project beneficiaries should not treat them as uni-dimensional stakeholders and overlook each person's diversity or meaning of life.

Issue 2: Insufficient attention to medium- and long-term processes of reconstruction (insufficient flexibility of resettlement programs)

In large projects there is a discrepancy between two timeframes: the resettlement program and actual time needed for genuine and full livelihood reconstruction. This discrepancy suggests that it may be difficult to produce results that will simultaneously satisfy both the project owners and resettlers. For example, in resettlement practices assisted by aid agencies, compensation and other inputs for livelihood reconstruction are usually made during actual relocation and the next several years, while it may take 10, 20, or even 30 years for resettlers to reconstruct their livelihoods or to establish a stable livelihood for the next generation. During the course of livelihood reconstruction, resettlers may have to face unexpected events or suddenly changing socioeconomic environments. Inputs given at the time of displacement may not be of any use later, in unanticipated situations. Therefore compensation and resettlement programs prepared based on the IRR Model or risk assessment beforehand may not give resettlers a strong hand in an uncertain future. It is obviously impossible to prepare for every risk identified in a risk assessment. Compensation and resettlement programs designed before displacement sacrifice flexibility to deal with future uncertainty, even if donor assistance was conditional on those programs.

De Wet points out that the complexity of displacement and resettlement are usually underestimated: "Certain aspects of the resettlement process do not seem to be readily amenable to the essentially rational, technical planning type of approach preferred by officials." De Wet says this is because "[e]verything happens all at once, a whole range of things, of different orders" (2006, p. 187). At the same time, de Wet recognizes: "In a number of cases, resettlers have taken initiatives and/or capitalized on unplanned for opportunities, thereby improving their situation" (2006, p. 181). These observations bring him to conclude that compensation and resettlement programs need to be open-ended and flexible (de Wet 2006, p. 199). Our research program is also based on the same perception that complexity and uncertainty may not necessarily be overcome by rational planning. The IRR Model, which enables us to factorize complex processes of impoverishment, has theoretical value. If planners depend solely on this model in preparing compensation and resettlement programs, however, they may encounter other difficulties to cope with the complex reality and uncertainty in the livelihood reconstruction process.

In sum, two theories and current practices based on them have both positive and negative aspects. The more we pursue merits by increasing the "sophistication," the more we may lose "respect" for resettled people and for complexity in resettlement processes (de Wet 2006, p. 199). Our research program started by recognizing these inherent or persistent difficulties in displacement and resettlement caused by dam developments. Empirical evidence in longitudinal studies may provide useful lessons to improve conceptual frameworks of compensation and resettlement programs, and may improve decision-making in development projects that involve displacement of populations.

Aim of this volume and description of chapters

Based on the critical review of policies and practices described above, we conducted field surveys from 2006 to 2014 on resettlement programs implemented for 17 large dam-construction projects in Asian countries. The field surveys are summarized in Table 1.1 and on the map that appears in the prelims. With the exception of two dams (Nam Theun 2 and Yali Falls), resettlement was completed more than two decades ago. Most of our findings were published as two special issues of the *International Journal of Water Resources Development* in 2009 and 2013. Chapters 2 to 5 include edited versions of articles in the special issues of that journal, with permission from the authors and publisher. Original articles are shown in Table 1.2. We allocated each article (case study) to the most relevant chapter. Each article presents various issues, which we analyze comprehensively and discuss in Chapter 7.

Chapter 2 presents cases of Ikawa, Nam Ngum 1, Wonorejo, Saguling, Kusaki, and Jintsugawa, and demonstrates issues at the planning stage of

Table 1.1 Large dam construction projects where resettlement was investigated

Country	Dam project	Period of resettlement
Indonesia	Bili-Bili	1990s
	Koto Panjang	1990s
	Saguling	1980s
	Wonorejo	1980s–1990s
Japan	Ikawa	1950s
	Jintsugawa	1950s
	Kusaki	1970s
	Miboro	1950s
	Miyagase	1980s
	Sameura	1960s–1970s
	Tokuyama	1980s
Lao PDR	Nam Ngum 1	1960s–1970s
	Nam Theun 2	2000s
Sri Lanka	Kotmale	1970s–1980s
Turkey	Atatürk	1980s
Vietnam	Song Hin	1990s
	Yali Falls	1990s–2000s

Table 1.2 Articles edited in this volume

International Journal of Water Resources Development	Chapter
Vol. 25, No. 3, 2009	
Naruhiko Takesada, Japanese experience of involuntary resettlement: long-term consequences of resettlement for the construction of the Ikawa Dam, 419–430.	2
Mikiyasu Nakayama and Kumi Furuyashiki, Renting submerged land for sustainable livelihood rehabilitation of resettled families, 431–440.	2
Atsushi Hattori and Ryo Fujikura, Estimating the indirect costs of resettlement due to dam construction: a Japanese case study, 441–458.	5
Syafruddin Karimi, Mikiyasu Nakayama, and Naruhiko Takesada, Condition of poverty in Koto Panjang resettlement villages of West Sumatra: analysis by using survey data of families receiving cash compensation, 459–466.	3

(continued)

Table 1.2 (continued)

International Journal of Water Resources Development	Chapter
Rampisela Dorotea Agnes, Mochtar S. Solle, Adri Said, and Ryo Fujikura, Effects of construction of the Bili-Bili Dam (Indonesia) on living conditions of former residents and their patterns of resettlement and return, 467–478.	3
Jagath Manatunge, Naruhiko Takesada, and Lakshman Herath, Livelihood rebuilding of dam-affected communities: case studies from Sri Lanka and Indonesia, 479–490.	2

Vol. 29, No. 1, 2013

Dian Sisinggih, Sri Wahyuni, and Pitojo Tri Juwono, The resettlement programme of the Wonorejo Dam project in Tulungagung, Indonesia: the perceptions of former residents, 14–24.	2
Sunardi, Budhi Gunawan, Jagath Manatunge, and Fifi Dwi Pratiwi, Livelihood status of resettlers affected by the Saguling Dam project, 25 years after inundation, 25–34.	2
Syafruddin Karimi and Werry Darta Taifur, Resettlement and development: a survey of two of Indonesia's Koto Panjang resettlement villages, 35–49.	4
Hidemi Yoshida, Rampisela Dorotea Agnes, Mochtar Solle, and Muh. Jayadi, A long-term evaluation of families affected by the Bili-Bili Dam development resettlement project in South Sulawesi, Indonesia, 50–58.	5
Bounsouk Souksavath and Miko Maekawa, The livelihood reconstruction of resettlers from the Nam Ngum 1 hydropower project in Laos, 59–70.	2
Bounsouk Souksavath and Mikiyasu Nakayama, Reconstruction of the livelihood of resettlers from the Nam Theun 2 hydropower project in Laos, 71–86.	3
Jagath Manatunge and Naruhiko Takesada, Long-term perceptions of project-affected persons: a case study of the Kotmale Dam in Sri Lanka, 87–100.	3
Erhan Akca, Ryo Fujikura, and Cigdem Sabbag, Atatürk Dam resettlement process: increased disparity resulting from insufficient financial compensation, 101–108.	5
Kyoko Matsumoto, Yu Mizuno and Erika Onagi, The long-term implications of compensation schemes for community rehabilitation: the Kusaki and Sameura dam projects in Japan, 109–119.	5

resettlement. Resettlers at Ikawa were provided with a new resettlement village, expecting they would benefit from rice cultivation. However, the economy of the village did not develop as expected because the quality of the rice produced was not good enough for the rice market. Resettlement of Nam Ngum 1 presents an exceptional case. No resettlement program was planned as the project was implemented during the Laotian Civil War, and much room was left regarding infrastructure development. In the case at Wonorejo, the dissatisfaction of resettlers could be attributed to insufficient road construction planning. At Saguling, resettlers were supposed to benefit from aquaculture at the reservoir, but they did not due to a lack of necessary legislation. This chapter also introduces two examples of good practices that contributed to the satisfaction of the resettlers. Resettlers at Kusaki participated in planning from a very early stage. Resettlement at Jintsugawa adopted a "rent scheme" because of time constraints for negotiation, but it resulted in a high level of satisfaction for resettlers.

Chapter 3 presents obstacles hindering proper resettlement implementation, discussing the Bili-Bili, Koto Panjang, Nam Theun 2, Kotmale, and Samueura dams. Resettlement areas for both of the two Indonesian cases, the Bili-Bili and Koto Pangjang dams, were not developed as promised when resettlers arrived at new locations, resulting in hardship for them. Sufficient land was not secured for the resettlers from Nam Thuen 2, and the farmland provided was not enough to support resettlers' livelihoods. In the case of Kotmale, an unexpected event (a decrease in the international market price of rice) dissatisfied the resettlers, who had been provided with paddy fields as land-based compensation. In the case of Sameura, the participation of the resettlers' community in planning was delayed because of strong opposition by residents against the dam construction, so the dam developer negotiated compensation individually with affected households. As a result, the community lost not only its unity but also an opportunity to obtain satisfactory compensation for submerged public facilities.

Chapter 4 demonstrates the positive effects of development by bringing secondary income sources to stabilize and increase resettlers' income, discussing the Koto Panjang, Nam Thuen 2, Yali Falls, and Song Hin cases. In the case of Koto Panjang, each submerged village was resettled as a whole. Villages that started fish breeding and sales in addition to agriculture succeeded in increasing and stabilizing villagers' incomes. Income differences between such villages and others increased. In Nam Thuen 2, a comparative study of two resettlement villages revealed that secondary incomes from handcraft production played a crucial role in significantly increasing resettlers' incomes. In Vietnam, resettlers from Yali Falls who diversified crops after resettlement were able to increase their income more than those from Song Hin.

Chapter 5 describes the emotional attachments of the resettlers to their submerged original homeland, presenting the cases of Bili-Bili, Atatürk, and Miyagase. In order to implement the resettlement program for Bili-Bili, resettlement areas were developed some hundred kilometers away from the submerged area, but they were not properly developed, as described in Chapter 3. After resettlement, many resettlers returned to the areas near the reservoir. One of the major reasons was their emotional attachment to their old homeland. Resettlers from Atatürk dispersed from around the submerged areas. Consequently, the resettlers lost the opportunity to visit their relatives often, though it had been a common activity among them, and large land owners lost their social prestige by only having status as newcomers in resettlement areas. Despite generous compensation, resettlers from Miyagase still feel they were sacrificed for national economic development.

Chapter 6 presents the living conditions of those who stopped farming after resettlement from Miboro, Tokuyama, and Bili-Bili. Many resettlers from Miboro moved to Tokyo and other urban areas, far from their homes in the mountains. They invested the cash received as compensation to purchase bathhouses or small hotels for couples. They chose these occupations because they had not been given vocational training at the time of resettlement, and they believed that these businesses could be managed without sophisticated skills. They generally enjoyed high incomes after relocation. Resettlement from the Tokuyama Dam took place in the late 1980s, when Japan as a whole was in an economic boom. It was not difficult for the resettlers to find jobs in urban areas even without any particular training or experience. Most of the resettlers from Bili-Bili who moved to urban areas eventually stopped farming. Those continuing farming did not depend on their farm income. Many of them were already retired and cared for by their children. Most of the second-generation resettlers graduated from high-school or university, and are satisfied with life after relocation.

In Chapter 7, we provide a comprehensive analysis of the results presented in previous chapters, and discuss measures necessary for future resettlement in Asia.

Acknowledgments

We express our sincere appreciation to Professor Michel Cernea for his constructive comments and suggestions for this book. The maps depicting the locations of dams and resettlement areas in the prelims and in Figure 3.2 were created by Osamu Furuta. This work was supported by JSPS KAKENHI Grant Numbers 18310033 and 24310189, Mitsui & Co., Ltd Environmental Fund, and the Hosei Society of Humanity and Environment.

Notes

1 In 2010, the Indian government admitted that 51,000 families would be affected by the development project (*The Hindu* 2010).
2 There is a strong account asserting that the World Bank absorbed such criticism as a "radical-movementist view" to sustain its hegemonic power by mainstreaming the countermeasures in development practice (Goldman 2005).

3 The concept of the IRR Model is explained in the various literature. One well-elaborated version is found in Cernea (2000).

4 For example, a planner may tend to think that resettlers would choose to resettle to a distant resettlement site if the government prepared a large area of irrigated farmland there. In the past, a planner may have tried to avoid moving people far away from the original place in order to reduce the stress of resettling (Cernea 1993, p. 25). But nowadays, a planner may tend to put more confidence in the economic rationality of resettlers when moving them to a distant site. Evidence from the field, however, indicates that resettlers do not always follow the economic rationality expected by planners.

References

Biswas, A. K. (2004). Dams: cornucopia or disaster?, *International Journal of Water Resources Development*, 20 (1), 3–14.

Bradlow, D. (2001). The World Commission on Dams' contribution to the broader debate on development decision-making, *American University International Law Review*, 16, 1531–1572.

Cernea, M. (1993). Anthropological and sociological research for policy development on population resettlement. In Cernea, M. and Guggenheim, E. (eds.) *Anthropological Approaches to Resettlement Policy, Practice and Theory*, Boulder, CO: Westview Press, pp. 13–38.

Cernea, M. (2000). Risks, safeguards, and reconstruction: a model for population displacement and resettlement. In Cernea, M. and McDowell, C. (eds.) *Risks and Reconstruction: Experiences of Resettlers and Refugees*, Washington, DC: World Bank, pp. 11–55.

Cernea, M. (2003). For a new economics of resettlement: a sociological critique of the compensation principle, *International Social Science Journal*, 175, 37–45.

Cernea, M., Guoqing, S., Hazar, T., and Kir, T. (2008). *Institutions and Capacity Building for Resettlement in Ilisu, Report on the Second Field Visit of the Committee of Experts: Resettlement March 10th–March 19th, 2008*. Retrieved from www.agaportal.de/pdf/nachhaltigkeit/coe_resettlement_20080613.pdf.

de Wet, C. (2006). Risk, complexity and local initiative in forced resettlement outcomes. In de Wet, C. (ed.) *Development-Induced Displacement: Problems, Policies and People (Studies in Forced Migration, Volume. 18)*, Oxford: Berghahn Books, pp. 180–202.

Dwivedi, R. (1999). Displacement, risks and resistance: local perceptions and actions in the Sardar Sarovar, *Development and Change*, 30, 43–78.

Dwivedi, R. (2002). Models and methods in development-induced displacement (review article), *Development and Change*, 33 (4), 709–732.

Fujikura, R. and Nakayama, M. (2002). Study on feasibility of the WCD guidelines as an operational instrument, *International Journal of Water Resources Development*, 18 (2), 301–314.

Goldman, M. (2005). *Imperial Nature: The World Bank and Struggles for Social Justice in the Age of Globalization*, New Haven and London: Yale University Press.

Goulet, D. (2005). Global governance, dam conflicts, and participation, *Human Rights Quarterly*, 27, 881–907.

Hirose, T. (2003). Manshu koku ni okeru Mizutoyo Dam kensetsu (The construction of the Sup'ung Dam in Manchukuo), *Niigata Kokusai Daigaku Jouhou Bunka*

Gakubu Kiyou (Bulletin of Niigata University of International and Information Studies Department of Information Culture), 6, 1–25.

The Hindu (2010). Sardar Sarovar: 40,000 families still to be resettled, *The Hindu*, October 25, 2010. Retrieved from www.thehindu.com/news/national/sardar-sarovar-40000-families-still-to-be-resettled/article847281.ece.

McDonald, M. J. and Muldowny, J. (1982). *TVA and the Dispossessed*, Knoxville: The University of Tennessee Press.

McDowell, C. (ed.) (1996). *Understanding Impoverishment: The Consequences of Development-Induced Displacement (Refugee and Forced Migration Studies, Volume 2)*, Providence, RI and Oxford: Berghahn Books.

Picciotto, R., Wicklin, W., and Rice, E. (eds.) (2001). *Involuntary Resettlement Comparative Perspective (World Bank Series on Evaluation and Development, Volume 2)*, New Brunswick, NJ: Transaction Publishers.

Quarles van Ufford, P., Giri, A. K., and Mosse, D. (2003). Interventions in development: towards a new moral understanding of our experiences and an agenda for the future. In Quarles van Ufford, P. and Giri, A. K. (eds.) *A Moral Critique of Development: In Search of Global Responsibility*, European Inter-university Development Opportunities Study Group Series, New York: Routledge, pp. 3–40.

Scudder, T. (2005). *The Future of Large Dams: Dealing with Social, Environmental, Institutional and Political Costs*, London: Earthscan.

Scudder, T. (2012). Resettlement outcomes of large dams (Chapter 3). In Tortajada, C., Altoinbilek, D., and Biswas, A. K. (eds.) *Impact of Large Dams: A Global Assessment*, Berlin: Springer, pp. 37–67.

Viegas, P. (1992). The Hirakud Dam oustees: thirty years after. In Thukral E. G. (ed.) *Big Dams, Displaced People: Rivers of Sorrow Rivers of Change*, New Delhi: Sage Publications, pp. 29–53.

WCD (2000). *Dams and Development: A New Framework for Decision-Making*, London: Earthscan.

World Bank (1994). *Resettlement and Development: The Bankwide Review of Projects Involving Involuntary Resettlement 1986–1993*, Washington, DC: Environment Department, World Bank.

World Bank (2004). *Involuntary Resettlement Sourcebook: Planning and Implementation in Development Projects*, Washington, DC: World Bank.

Box 1.1 Demand for electricity and water

Demand for electricity and water in the Asian region has been increasing rapidly, along with economic development and improvement of life in the region. Table 1.3 presents per capita consumption of electricity. Consumption in 2010 in Indonesia (641 kWh) and Sri Lanka (445 kWh) was far less than in Japan (8,339 kWh) or in Organisation for Economic Co-operation and Development (OECD) member countries (8,315 kWh). Even Turkey (2,474 kWh) consumed less than China (2,958 kWh). It is likely that the demand for electricity in Asia will increase further.

Table 1.3 Electricity consumption and output of selected countries in 2010

Units	Consumption	Output		
		Total	Hydro	Hydro share of total
	kWh per capita	*TWh*	*TWh*	*%*
World	2,892	21,431	3,437	16.0
OECD total	8,315	10,854	1,351	12.4
Asia (excluding China)	806	2,078	259	12.4
China	2,958	4,247	722	17.0
Indonesia	641	170	18	10.4
Japan	8,399	1,111	82	7.4
Sri Lanka	445	11	6	52.3
Turkey	2,474	211	52	24.5
Vietnam	1,035	95	28	29.0

Sources: IEA (2011b, 2011c).

The electrification rate in Indonesia in 2009 was 64.5 percent, well below that of East Asia (90.8 percent), and 81.6 million people in Indonesia had no access to electricity in 2009 (IEA 2011a). Indonesia had a low dependence on hydropower (Table 1.3), with only 5,705 MW developed so far (Harris 2012). Hydropower potential in Indonesia is estimated at 75,670 MW, however, making it one of the world's top ten countries for hydropower potential (IEA 2010). Hydropower development therefore appears to be a feasible option to meet electricity demand in Indonesia.

In the Asia and Pacific region, only 18 percent of technically exploitable hydropower potential has actually been exploited (IEA 2010). It is estimated that hydropower capacity in the developing Asian region, excluding China and India, will expand from 40 GW in 2010 to 134 GW in 2040, an annual average growth rate of 4.1 percent (USIEA 2013).

As economies develop, people tend to consume more food produced by animals. More grains must be produced to feed more animals, to meet the increasing demand as people improve their quality of life. Among the Asian countries whose dam projects are covered in this book, all countries except Japan have increased cereal production by an annual rate of between 2.1 percent and 4.1 percent during the last half century, as shown in Table 1.4. (Japan has been heavily dependent on imported foods—feed grains, in particular.) During the same period, irrigated land has also expanded (Table 1.5).

(continued)

(continued)

Table 1.4 Cereal production of selected countries

Year	Cereal production (million tons)					
	Indonesia	Japan	Lao PDR	Sri Lanka	Turkey	Vietnam
1961	14.4	16.5	0.6	1.0	12.7	9.3
1971	22.8	11.9	0.8	1.4	20.9	10.7
1981	37.3	11.2	1.2	2.3	25.5	12.8
1991	50.9	10.6	1.3	2.4	31.1	20.3
2001	59.8	9.9	2.4	2.7	29.6	34.3
2011	83.4	11.5	4.2	4.0	35.2	47.2
Average annual change 1961–2011 (%)	3.6	−0.7	4.1	2.9	2.1	3.3

Source: http://faostat3.fao.org/home/E.

Table 1.5 Irrigated land of selected countries

Year	Total area equipped for irrigation (1000 ha)					
	Indonesia	Japan	Lao PDR	Sri Lanka	Turkey	Vietnam
1961	3,900	2,940	12	335	1,310	1,000
1971	3,900	3,364	19	439	1,850	1,200
1981	4,107	3,031	116	500	2,835	1,800
1991	4,410	2,825	140	530	4,100	2,900
2001	5,745	2,624	300	570	4,985	3,850
2011	6,722	2,474	310	570	5,215	4,600
Annual average change 1961–2011 (%)	1.1	−0.3	6.7	1.1	2.8	3.1

Source: http://faostat3.fao.org/home/E.

Chartres and Varma (2011) estimated future water demand and predicted that another 2,500–6,000 km^2 of water resources will need to be developed by 2050, assuming a global population of nine billion in 2050 and unchanged agricultural productivity. The predicted future demand for water resources is equivalent to another 25 to 50 enormous dams like the Aswan High Dam on the Nile River in Egypt.

Considering the increasing demand for electricity and water in the Asian region, it is fair to say that more dams, large or otherwise, are likely to be planned and constructed in the future in order to supply both electricity and water.

References

Chartres, C. and Varma, S. (2011). *Out of Water: From Abundance to Scarcity and How to Solve the World's Water Problems*, Upper Saddle River, NJ: FT Press.

Harris M. (2012). *PLN Report Shows Indonesia Has Potential for Significant Hydroelectric Growth*, Renewable Energy World, January 25. Retrieved from www.renewableenergyworld.com/rea/news/article/2012/01/pln-report-shows-indonesia-has-potential-for-significant-hydroelectric-growth.

IEA (International Energy Agency) (2010). *Renewable Energy Essentials: Hydropower*. Retrieved from www.iea.org.

IEA (2011a). *World Energy Outlook 2011*. Paris: IEA.

IEA (2011b). *Energy Balance of Non-OECD Countries (2011 Edition)*. Paris: IEA.

IEA (2011c). *Energy Balance of OECD Countries (2011 Edition)*. Paris: IEA.

USEIA (US Energy Information Association) (2013). *International Energy Outlook 2013*. Washington, DC: USEIA.

(Ryo Fujikura)

2 Planning resettlement programs

Introduction

Resettlement programs must be carefully designed and implemented so as to be an opportunity for people to become better-off, but there have been many cases where this was not so. Even if the program was carefully prepared, the full consequences of resettlement may not be as expected. In this chapter, we review six cases of dam resettlement programs: three in Japan, two in Indonesia, and one in Lao People's Democratic Republic (PDR). First, we present two cases of land-for-land compensation: the Ikawa Dam in Japan and Nam Ngum 1 Dam in Lao PDR. The former was constructed half a century ago, before the Japanese government adopted a cash compensation policy. The latter was constructed in 1971 during the county's civil war. The Lao project neither involved resettlers in project planning nor development of a resettlement program. Then, we present the two cases of Wonorejo and Saguling in Indonesia, where compensation included cash and participation in the Transmigration Program (TP), which was introduced to facilitate immigration from the densely populated islands of Java and Bali to less populated ones such as Sumatra and Sulawesi. In both cases, the majority of resettlers chose to move to areas near the reservoir, while the government had assumed that many people would choose to move to more remote areas, including other islands. As a result, the natural resources in the resettlement area became insufficient to improve the lives of resettlers. Finally, we present two Japanese cases in which resettlers were relatively satisfied with resettlement. In the Kusaki Dam case, where compensation was principally made by cash, resettlers were able to negotiate with the developer in order to continue their jobs after resettlement. In the case of the Jintsugawa Dams, a unique rent scheme was employed for the resettlers to maintain their livelihoods and their pride as landowners.

Ikawa Dam (Japan)

The Ikawa Dam project, completed in 1957, was one of the few attempts to devise a good compensation practice by adopting land-for-land compensation.

This was before adoption of the Japanese compensation guidelines, which stipulate only monetary means for compensation (see Box 2.1 at the end of this chapter). In this remote mountainous area, at about 650 meters elevation, people were mainly engaged in small-scale farming and forestry. Their staple food was not rice, but millet harvested through slash-and-burn agriculture on the mountain. Since there were no roads for automobiles connecting the village of Ikawa with the outside, goods for everyday life were carried by people crossing the nearest pass. There were no major industries other than forestry, which utilized the rivers to transport logs. There were two elementary schools and one junior high school in the village. Most of the students who graduated from junior high school did not go to high school in downstream areas, such as the city of Shizuoka, but remained in the village to become part of the workforce for their households.

After rigorous consultations among parties, in 1953 Chubu Electric Power Company finally agreed to the three conditions of compensation demanded by the villagers: (1) completion of the Dainichi Road (to Shizuoka); (2) construction of a new village; and (3) full and satisfactory compensation for a better and improved living standard (Ikawa Village 1958). One of the major requests of the villagers was the construction of the "New Village." At the same time, there was an option to receive cash compensation instead, and eventually 99 out of the 193 resettlers took the cash compensation option to leave the village.

This compensation scheme, known as New Village Building, aimed to provide compensation on a land-for-land basis. Some resettlers who had lost their houses and cultivated land received new housing and new land plots in a newly developed area within the village. Others received new reclaimed land near the reservoir for their housing within the original main village, which was also equipped with new infrastructure and community facilities. In cases where the property could not be replaced with substitutes, cash compensation was given. In one such newly developed area, known as Nishiyama-daira, near the main village, 23 houses were built with attached land plots and other community facilities, including water and electricity supply.

There were two special features in this compensation scheme. First, in the newly developed land, rice cultivation was newly introduced. At a high elevation of around 650 meters, villagers in Ikawa did not have any substantial paddy cultivation before the inundation for the reservoir. It was said that many villagers had only known rice as a precious commodity that was eaten just once or twice a year, but did not know much about the rice plant itself. Second, for this new practice, one agricultural expert from Shizuoka Prefecture was stationed for four years in Nishiyama-daira in order to assist villagers in stabilizing their agricultural production and thus improving their livelihoods. This expert conducted a test on the cultivation of rice in the newly developed Nishiyama-daira one year before the relocation in order to obtain confidence from the villagers.

Several factors enabled this land-for-land compensation policy. First, before the construction of the Ikawa Dam, forced relocation with cash compensation was already perceived as a risky practice, especially for the farmers living in remote rural areas, because they lacked the experience in handling such huge amounts of cash. When the compensation was paid by installments, the amount of cash was insufficient to support their life during periods of inflation. There was a number of cases where resettlers failed to invest in land large enough to rehabilitate their lives, due to an increase in land prices stimulated by the compensation cash itself. Resettlers were forced to borrow money to come up with the necessary funds to invest in land or other businesses, and the debt became an additional financial burden for them (Takashima 1956; Hanayama 1969). Therefore, land-for-land compensation was thought to be a better option, even though it was more expensive according to the company's budget estimate compared to cash compensation in the Ikawa case (Chubu Electric Power Company 1961). Second, the Chubu Electric Power Company was desperately eager to construct the Ikawa Dam and power plant to meet the rapidly growing electricity demand in that period. This consideration on the part of the power company was the main reason for introducing the New Village Building policy. Third, the Shizuoka prefectural government played an active role in negotiation, preparation, and implementation of the resettlement program (Takesada 2006). In those days, it was responsible for the management and development of the major river in its territory, so it was a genuine stakeholder of the Ikawa Dam construction project. Thus, the Shizuoka prefectural government tried to arbitrate the negotiations between Ikawa villagers and the Chubu Electric Power Company, and prepared a resettlement program, along with a feasibility study and technical assistance from an agricultural expert. Fourth, the villagers were ready to negotiate when the plan was made public. As mentioned earlier, the plan for the Ikawa Dam had been long standing, and villagers knew that, sooner or later, the dam would be constructed. One of the committee members interviewed by the author said that since his childhood there had been several surveys in the area, and he had thought it would be in the destiny of Ikawa to have a dam someday.

The short history of the Ikawa area after the dam construction is a typical case of depopulation in rural Japan in the 1960s and thereafter. During the first ten years after the dam was completed (1957–1966), there was prosperity and hope in the village. Construction of two other dams and one power plant in the upstream area provided villagers with job opportunities as well as commercial opportunities in the area. In 1962, the population in the Ikawa area was more than 8,000. With completion, in 1958, of Dainichi Road (the road to the city of Shizuoka)—one of the villagers' conditions for agreeing to the dam construction—bus transportation from the city commenced, and many electrical appliances and consumer goods flooded into the village. Tourism development utilizing the newly created reservoir was expected to boom. The next ten years (1967–1976), however, was a period when the depopulation and marginalization of Ikawa began. In this decade, the population of Ikawa fell to lower than that prior to construction of the Ikawa Dam. The village of Ikawa was incorporated into the Shizuoka City Municipality, and the tax revenue from the dams and power plants was absorbed into the city's accounts. The quantity and quality of agricultural production in the resettled land was not enough to merchandize products outside the village. Rice cultivation was gradually abandoned under the pressure of the national policy of reduction of rice acreage. Moreover, even tourism could not become a major industry without attractions such as hot springs. Since 1977, Ikawa has been suffering from not only depopulation but also an aging population. By the year 2000, the population of Ikawa had decreased to about 760, and the ratio of population older than 65 years of age exceeded 50 percent. There have been several plans and activities for revitalizing the area by promoting new agricultural production or attracting tourists with new events, but none of them has succeeded in shifting the trend. Many inhabitants of the Ikawa area now strongly desire a tunnel that bypasses the current mountainous road to Shizuoka, so that their children can commute either to school or to work from their houses in Ikawa. In other words, people are unable to think of any other effective alternatives to alter the course of Ikawa's future.

Interviews with inhabitants of Nishiyama-daira were conducted from January to March 2006, and 19 of the 24 households were covered. Interview results regarding the resettlement program are summarized as follows:

1   The decision to move to Nishiyama-daira was made individually. From the former hamlet called Shimawago, all 19 households moved to Nishiyama-daira, but none of the interviewees said that the decision was collectively taken in the hamlet, with each household instead considering its own interests and constraints. For example, to start with some of them did not even think of leaving the village at all, while others reluctantly obeyed their elders' decision.

2   The land-for-land compensation scheme was evaluated favorably. The extent of satisfaction and/or dissatisfaction with compensation varied among the households. Some of them expressed their appreciation to the Chubu Electric Power Company, while others expressed their dissatisfaction when compared with the recent dam compensation in the nearest area downstream of the Ikawa Dam. However, appreciation for land-for-land compensation was generally shared.

3   The most significant change after dam construction was the improvement of road connections. Ikawa did not previously have an automobile road connection with the city of Shizuoka, about 50 kilometers away. Reaching the city took more than six hours, including three hours for crossing the nearest pass on foot. The completion of the Dainichi Road in 1958 changed not only the material life of Ikawa residents but also

people's mind-set and lifestyle. As the number of passengers and materials coming into the village increased, the villagers recognized the necessity of cash income. They also obtained information on occupations and life-styles that were novel to them, and leaving the village for urban areas became one of the feasible options.

4   The newly introduced rice cultivation was welcomed for a while. All the resettlers actively engaged in rice farming for several years, but the quantity and quality of rice was not sufficient to be sold outside the village. As mentioned earlier, in the 1970s most of the resettlers stopped rice farming. At present, only two households continue this practice. At the planning stage, Shizuoka Prefecture made various efforts to encourage villagers to engage in rice farming in Nishiyama-daira. Although the resettlers engaged in rice farming, as expected, their perception and level of interest was different from what the Shizuoka Prefecture officials had expected. In the interview, only a few of the resettlers expressed their favorable appreciation for the agricultural expert who was stationed in Nishiyama-daira for four years.

5   The present difficult situation of Ikawa is another story from the dam construction. The resettlers did not think that the current situation of Ikawa had been brought about by the dam and resettlement. They did not consider themselves to be victims of development, and did not consider benefits for the downstream areas. They considered the dam as a part of change that had both merits and demerits.

Resettlement as a new opportunity

Based on these interview results, the question of whether or not resettlement to Nishiyama-daira was a new opportunity for resettlers can be answered as follows. First, at the level of perception, there were two parties concerned. One view is that of the power company and prefectural government as planners of resettlement. The other view is that of the resettlers. From the former view, resettlement to Nishiyama-daira was definitely a new opportunity given to villagers, especially with its agricultural promotion program, which included rice farming. As part of the plan, it was envisaged that a new Ikawa village would become an example of an innovative agricultural area among mountainous areas at high elevations. From the latter view, things were not so simple. There were some differences in opinion among villagers with respect to their perception of risks and/or incentives. Some villagers perceived rice farming as a risky plan. A few took it as an incentive and engaged in agricultural activities vigorously after resettlement. The majority of the villagers felt that rice farming was risky, but that land provision was the most conservative and feasible option. For this reason, the resettlers chose to move to Nishiyama-daira. This shows that there was a gap between planners and resettlers in their perceptions of the resettlement options, and that there were differences among the resettlers themselves in their perceptions and choices.

Second, at the level of actual livelihood, the results are mixed. On the one hand, agricultural development did not take place as expected. Rice cultivation was also abandoned under the pressure of a changed national policy. In this sense, the new opportunity to develop agriculture as the main occupation for the new village was not realized, and the planners could consider it as a failed resettlement program. On the other hand, a new lifestyle, especially for the resettlers' children, was materialized with the help of higher education. In addition, forestry took advantage of improved transport connections by road, and the resettlers derived unexpected benefits from unutilized resources, although this was short-lived. Both developments were not clearly envisaged in the Ikawa resettlement program. In sum, although the new opportunity was realized in the actual livelihoods of resettlers in an unexpected way, the resettlers were still satisfied with it. After reviewing two aspects of the Ikawa resettlement—one regarding the perception or interpretation of the proposed resettlement plan; the other regarding the evaluation of the actual result of resettlement—it was noticed that these two aspects were not directly linked. Although the resettlers did not appreciate the resettlement plan as expected by the planner, they appreciated the consequences of the resettlement.

Conclusions

In this study, the experience of involuntary resettlement in Japan was reviewed with a focus on resettlers' choices and their long-term consequences. There are several findings from this study of the Ikawa Dam resettlement case. First, it can be concluded that the resettlers' motives, which lay behind their choices regarding the move to a newly developed resettlement area (Nishiyama-daira), were diverse. At first sight, the decision appeared to be a collective and unanimous one among all households in the former hamlet, as the planners of the resettlement program had expected. However, the interview results revealed that the decision was taken by each household independently, after considering their capacities, preferences, and constraints. Second, the resettlement program (New Village Building) was perceived differently by the planners and resettlers. The resettlers did not see the resettlement program that emphasized agricultural development, including new rice farming, as a new opportunity in the way planners had expected. However, this perception gap may have also mitigated the possible dissatisfaction of the resettlers, since the blueprint for the New Village Building was not realized. On the contrary, it seemed that since resettlers did not expect too much from the resettlement program, in the long run they felt satisfied with their livelihoods in the resettlement area. Last, the satisfaction of the resettlers had its root in the successful rearing of their children and in securing an independent livelihood for them in the city. After the resettlement, almost all the resettlers with children consciously provided them a better education, even if it involved incurring high costs for their children to stay in the city and leaving the Ikawa area. Neither this source of satisfaction nor possible strategies on the part of resettlers had been intentionally planned or incorporated into the resettlement program.

Nam Ngum 1 (Lao PDR)

The Nam Ngum 1 (NN1) Dam, built in 1971, was the first large dam constructed in Laos. With a reservoir area of 370 km², it was constructed on the Nam Ngum, a tributary of the Mekong River, and is about 100 km from Vientiane to the north by road.

Approximately 23 villages were settled in the Nam Ngum region. Within these villages were 570 households and 3,242 people affected by the construction of the NN1 Dam, inundated about 2,840 ha of land and 1,840 ha of paddy field (Schaap 1974). As stated in the original NN1 report, no "social and environmental impact assessments" were conducted prior to construction of the dam. This may be due to the effect of the Laotian Civil War (1953–1975), the absence of environmental regulations at the time, and the fact that in those days people were less concerned with the social and environmental impacts, given their belief that the dam was important for the development of the country.

The resettlement programs were developed by the government without any consultation with the resettlers. Most of the residents in both villages of Pakcheng and Phonhang, among several major destinations of the resettlers, expressed that they did not have any choice in the resettlement, since all the plans depended on the resettlement policy of the government. NN1 was constructed in 1971, with the main phases of pre-construction and construction occurring during the Indochina War. The project had no social impact assessment, and no resettlement action plan. All resettlers were forced to leave their old villages in a short period of time, with some families evacuating along the river by boat before the closure of the cofferdam. The project did not provide cash for compensation and did not build houses for the resettlers; it simply provided the land and resources for use in agriculture, temple construction, cemetery, and other infrastructure development.

Research was conducted on the resettlement villages of Pakcheng and Phonhang in June 2010. These two villages were selected in order to compare the livelihood conditions between two resettlement villages, which were established in different periods, before and after the end of the civil war. One hundred households (50 each from Pakcheng and Phonhang) were interviewed. The focus group interviews used for this survey were comprised of first- and second-generation residents of the resettlement villages. First generation refers to family heads and their children who are over or around the age of 50 and who could remember life in the old villages (43 households in Pakcheng, 44 in Phonhang). Second generation refers to those born after the resettlement (7 households in Pakcheng, 6 in Phonhang).

The Pakcheng resettlement village was the result of a merger of four old villages in 1968: Na Luang, Konsui, Na Khea, and Na Leang. People were first resettled temporarily in the Thalat area for about three years, and were then moved to the present Pakcheng. The resettlement policy was planned and implemented by the government of Lao PDR. The present village profile consists of 161 households: 105 resettled from the old villages, while the remaining 56 are second-generation residents. Of the survey and interviews conducted with 50 households (about 24 percent of all households), 43 resettled from the old villages. The remaining 7 households are second-generation residents. The average size of the 50 households interviewed in this village is about 5.5 persons.

The Phonhang resettlement village was the result of a merger of two different old villages: Kengnoi and Na Luang. Residents from Na Luang were resettled to both Pakcheng and Phonhang, which are about one kilometer apart. Phonhang currently consists of 120 households: 100 resettled from the old villages, while the remaining 20 are second-generation residents. The survey and interviews were conducted with 50 households, 44 of which had resettled from the old villages. The remaining 6 households are second-generation residents. The average size of the 50 households interviewed in the village is about 5.3 persons.

There was no private-sector employment in the old villages, and many residents were self-employed farmers practicing traditional agriculture (e.g., paddy rice field, slash-and-burn agriculture, family garden). While many of the residents of the resettlement villages still work as self-employed farmers, about 36 percent of families in Pakcheng are working outside the village. The village is located only 10 km away from the provincial center, making it possible for young people to work there as public-sector employees, laborers, and in other service jobs. Meanwhile, most residents of Phonhang are farmers, and only about 10 percent are working outside the village.

A comparison of family income between the old and the new resettlement villages cannot be made, because the value of the currency has significantly changed and most villagers were dependent not on money but on natural resources during the period of the resettlement. These days, the annual average family income in Pakcheng and Phonhang is LAK 23,112,000 (USD 2,889) and LAK 10,380,000 (USD 1,298) per year, respectively. The incomes of both villages were higher than the Lao PDR rural poverty line of USD 850 per household per annum in 2009. While 10 percent of households of Pakcheng have a minimum family income of LAK 3,000,000 to 5,000,000, in Phonhang the percentage is 22 percent. Meanwhile, 18 percent of residents of Pakcheng fall into the middle-income bracket of LAK 16,000,000 to 20,000,000, in Phonhang the percentage is only 4 percent.

Before resettlement, much land was available and most households had a traditional irrigated paddy rice field. The average land use prior to resettlement was about 1.5 ha per household. In the present resettlement villages, the land used for paddy rice fields is about 1.3 ha and 1 ha per household in Phonhang and Pakcheng, respectively. After the resettlement, households in both Pakcheng and Phonhang have continued the tradition of having an irrigated paddy rice field, as rice is their main food. The difference was the quality of agricultural land and the sources of irrigation. Pakcheng is located along the Nam Ngum River, whereas Phonhang is located relatively further away from the river. The resettlement policy has provided land for paddy cultivation of about 0.5 ha per household and about 900 m² for a home plot. The land use became limited compared to what was available in the old villages.

Infrastructure, living conditions, and educational facilities in Pakcheng are generally better than in Phonhang. The old villages had traditional irrigation systems and an ample supply of water. On the one hand, Pakcheng had the opportunity to benefit from the Pakcheng Agriculture Project, implemented from 1980 to 1995, which provided an ample irrigation system pumping water from the Nam Ngum River. On the other hand, Phonhang, which was resettled later, in 1977, has a different irrigation system and does not have an ample water supply due to poor pumping facilities. The housing styles in Pakcheng and Phonhang are different too, mostly due to the differing family incomes. While many of the houses in Phonhang are built of wood and bamboo, and the roof is often made of grass, those in Pakcheng are built with brick, cement, and other construction materials. When comparing the present houses with those in the old villages, the residents of Pakcheng expressed the view that they had become better, but those in Phonhang felt they were worse. Pakcheng is located along the main road that leads to the district center, and Phonhang is located about one kilometer from the main road. Residents of Pakcheng are able to sell their products to the market in their village. They can both walk to the market or use bicycles and motorcycles to get there, while the residents of Phonhang have to travel to the market in Pakcheng. Phonhang only has a primary school, while there is both a primary school and a high school in Pakcheng. When children in Phonhang finish primary school, they have to move to Pakcheng to enroll in middle and high school.

Many residents of both villages thought that the present villages are worse than the old villages. However, 80 percent and 58 percent of the residents of Pakcheng and Phonhang, respectively, expressed satisfaction with their present life in the villages, and 62 percent and 48 percent of the residents of Pakcheng and Phonhang, respectively, are satisfied with their current job. Residents of both Pakcheng and Phonhang indicated that they would like to live in their villages for a long time, since they do not have any other options. Most residents of both Pakcheng and Phonhang believe that the places they live in are good for their children because of improved public infrastructure, including electricity, road access, schools, public health services, and water supply, among others.

The differences in residents' income and satisfaction between Pakcheng and Phonhang can be attributed to the differences in accessibility to the main road to Vientiane and the differences in irrigation facilities. Phonhang has only one access road connecting to Pakcheng, and residents of Phonhang must go through Pakcheng to go to Vientiane. The existing irrigation system of Phonhang is poor and villagers can currently cultivate agriculture only during the wet season, even though the land is suitable for agriculture all year round. While there are 85 ha of paddy rice field in Phonhang, the present irrigation system can only supply water to about 10 ha.

Despite the fact that the NN1 project did not provide cash compensation to the resettlers, it provided sufficient land and resources for them. As a result, they could continue to secure their livelihoods without any cash compensation.

Resettlement in Indonesia

Indonesia has implemented the TP since the colonial period (see Box 2.2 at the end of this chapter). It aims to facilitate the immigration of people from the densely populated islands of Java and Bali to other less populated islands such as Sumatra or Sulawesi. Immigrants are supposed to be provided with 2 ha of farmland and a house free of charge in the TP areas. The Indonesian government has integrated the TP scheme with resettlement caused by some dam construction projects. Resettlers from submerged areas in Java can participate in the TP in addition to getting cash compensation for their submerged properties. It is an opportunity for resettlers who were landless or owned only a small plot of land and were unable to purchase enough land in resettlement areas.

Nonetheless, resettlers tend to stay close to the reservoir area, their original domicile. The reasons include: (1) the strong ties among relatives and between the people and their homeland; (2) insufficiently prepared land reported for transmigration; and (3) a prevalence of transmigrated people still living at a subsistence level (IOE 1985). In the cases of the Wonorejo and Saguling dams, which we present below, significantly fewer resettlers chose to join the TP than the developers expected. The developer of the Saguling Dam assumed that some 79 percent of resettlers would be transmigrated into other areas, but only 2.4 percent (74 of 3,078 resettlers) actually joined the TP (Nakayama 1998). As a result, land for resettlers near the reservoir became insufficient to accommodate their needs.

Meanwhile, many of those who had chosen the TP from among the Wonorejo resettlers seemed to succeed in improving their livelihood. According to the respondents' knowledge of the condition of their neighborhood, 52 respondents (59 percent) reported a success story in terms of their achievements in the new area, while 3 (3 percent) believed that they did not succeed in the transmigration area. The remaining 33 respondents (38 percent) reported not having enough information on their neighborhood. Sometimes reunions of old villagers still take place, and transmigrated people visit their relatives in the villages surrounding the reservoir. This possibility might change the perception of the transmigration scheme as an option in the resettlement program, but it remains difficult to change people's emotional reasons for remaining nearby.

We present another two dam construction cases in the next chapter: Koto Pangjang in Sumatra and Bili-Bili in Sulawesi. Resettlement of Koto Pangjang was conducted per village: one entire village was resettled to another location. Resettlement for the Bili-Bili Dam was conducted differently: resettlers were able to participate in the TP if they so wished. It was an exceptional offer to the resettlers, because the TP is, in principle, exclusively provided to those resettled from Java and Bali.

Wonorejo Dam (Indonesia)

The Wonorejo Multipurpose Dam was constructed in the Brantas River Basin in East Java Province, Indonesia (Sinaro 2007). This region is about 130 km southwest of Surabaya, the capital city of East Java Province, and is about 15 km west of the Tulungagung city center. There were 1,414 households (7,144 individuals) living in the village of Wonorejo. The affected area and number of persons amounted to a submerged area of 210 ha with 668 households. Another 909 households on 450 ha of land lived on terrain that would be isolated by the reservoir (IPB 1985).

There were two scenarios for resettlement prepared by the government:

1   The villagers of Wonorejo (1,414 households) could be moved to a new area outside of Java Island (preferably the southern portion of Sumatra Province). In this scheme, resettlers would need to join the TP (see Box 2.2).
2   A partial resettlement program targeted only those living in the submerged area (668 households). This scenario would be conducted using the TP or relocation to nearby villages.

There also used to be an additional scenario: a swamp reclamation project in the southern part of Tulungagung. This project was once supposed to absorb many resettlers. It later turned out that the farmland newly reclaimed by the project could accommodate only 668 households. This figure was much smaller than initially estimated. Therefore, the government decided not to implement this swamp reclamation project. As a result, the resettlement program was composed of a combination of the first two scenarios (Nippon Koei 2002).

According to the survey carried out by the authority in charge of the project, 475 households decided to join the TP, 356 households moved to surrounding villages, and 165 households chose to resettle upstream after receiving their compensation. This latter group that chose the second scenario owned dry and agricultural land in addition to still having relatives in the project area. The upstream area included remote villages with poor access to roads and fresh water, no sanitation, and no electricity. The resettlers were warned of these difficulties.

The resettlement program attempted to, at the very least, sustain or improve the living standards of resettled individuals. Furthermore, the project constructed infrastructure at the resettlement sites but also recognized that the growing number of the settlers moving upstream meant an increased environmental load, leading to greater environmental deterioration.

A condition of Japan's official development assistance loan for the dam project included an interview with 20 resettled residents in October 2004 to investigate the current condition of various public facilities built related to the project, and to obtain their opinions concerning the resettlement process and their current lifestyles within the village (Okada 2004). A summary of the interview results is as follows:

1   The land acquisition and resettlement process involved no disputes with residents and proceeded smoothly from beginning to end.
2   Roads, school buildings, health centers, and other public facilities were constructed, and resettled residents are satisfied with their living environment.
3   Although resettled individuals suffered no economic hardships in the past as a result of farming, transporting materials, or supplying labor during construction of the dam, they have found it more difficult to find work since the project was completed and have become anxious about their potential income. In this respect, it would be beneficial for the government to provide vocational training and job opportunities.

As these comments show, while the residents were not negative about their resettlement, they indicated that the government should have given more thought to their situation several years down the line. Their hope was that the government would provide some form of vocational training as a means of improving future living standards.

Field investigations and interviews were undertaken in July 2011 to map the perception of individuals exposed to an involuntary resettlement program focused on former residents of the dam construction area. Eighty-eight respondents were randomly selected as a heterogeneous crowd, where different groups of people represented different age groups and both sexes. They included:

1   Those who held an independent household even before resettlement and who made the decision at the time of resettlement (they are considered as first generation).
2   Those who became independent at the time of resettlement (i.e., they received compensation separately from their parents) or thereafter. This group also includes those who did not make their own choice for resettlement, but obeyed the decision made by their parents (this group is considered second generation).

The respondents did not all participate in the resettlement scheme at the same time. At the initial stages, resettlement was conducted gradually and afforded first priority to those living in the submerged area; second priority was then given to residents near the construction site. Since there were financial problems with the dam construction when donor countries suspended their support, first-priority relocations only started in 1982/1983 and extended until 1992/1993 (first stage). In 1994/1995, the government committed national resources to continue the construction of the Wonorejo Dam. The resettlement program was re-initiated in 1994/1995 and 1995/1996 (second stage). Accordingly, the respondents were identified as members of the first stage (44 respondents) and members of the second stage (44 respondents).

Some main indicators of public perception are summarized in Table 2.1. Before accepting the resettlement program many of the respondents did not own land to engage in agriculture. Fifty-four (74 percent) respondents living upstream and 13 (87 percent) respondents living downstream indicated that they presently own more land. According to the resettlement priority that an individual was associated with, there may have been more or less land available. Similarly, 40 respondents (91 percent) from the first priority and 27 (61 percent) respondents from the second priority reported presently

Table 2.1 Summary of number of respondents concerning the conditions before and after the resettlement program, according to respondent classification

Main indicator	Relocation area		Relocation period		Generation	
	Upstream	*Downstream*	*First stage*	*Second stage*	*First generation*	*Second generation*
A Income stability						
Better	24	0	7	17	14	11
Worse	30	6	17	14	34	1
Other	10	2	6	6	10	2
No answer	9	7	14	7	12	4
B Land ownership						
Increase	54	13	40	27	58	9
Decrease	17	2	3	16	10	9
Other	0	0	0	0	0	0
No answer	2	0	1	1	2	0
C Property						
Increase	37	4	19	22	27	14
No change	23	5	16	12	25	2
Decrease	11	6	9	8	16	1
No answer	2	0	0	2	2	1

D Social community						
Better	63	14	39	39	61	16
Worse	8	1	5	4	7	2
Other	1	0	0	1	1	0
No answer	1	0	0	0	1	0
E Compensation scheme						
Better	45	10	30	26	47	7
Worse	6	4	8	2	7	3
Other	21	1	6	16	14	8
No answer	1	0	0	0	2	0
F General satisfaction						
Better	68	15	43	27	67	16
Worse	0	0	0	0	0	0
Other	0	0	0	0	0	0
No answer	5	0	1	17	3	2

Source: Authors.

owning more land for agriculture. The remaining resettled individuals struggled to find adequate land for agriculture.

Most respondents experienced instability with their income after entering the resettlement program, though 17 respondents from the second priority of resettled individuals (39 percent) expressed that their income was stable. This was thanks to better compensation and a better political situation in this period. The instability associated with the first-priority resettled individuals was linked to insufficient skills and age; competition for new jobs was difficult when they had only worked in agricultural sectors.

Except for income stability, there are almost no differences in the livelihood condition of respondents living in the upstream or downstream areas, as well as those who resettled in different years. There were advantages and disadvantages for both the upstream and downstream areas. Those who moved to the downstream area of the reservoir had access to better technology and good public facilities; however, their economic burden was heavier than those living in the upstream area. As a consequence, six respondents (40 percent) indicated that their property was reduced. In contrast, those living in the upstream area of the reservoir obtained additional income by temporarily utilizing communal forestland for agriculture. They now have enough land for cultivation despite the fact that the quality of land is worse when compared to what they owned in the past. This was one of the reasons that those living upstream indicated that their income was not stable.

Resettlers had strong emotional reasons for choosing to remain in the surrounding villages. Formerly, they were introduced to the options of either transmigration or moving to nearby villages. The government also promised those in the TP more land and greater cash compensation. Despite the available options, they chose to move to nearby villages for emotional reasons. Specifically, 63 respondents (59 percent) wanted to stay near the reservoir area, 25 respondents (23 percent) still occupied land that was not submerged, 7 respondents (7 percent) were not confident with the government's promise and their own capacity to survive in the transmigration area, and 12 respondents (11 percent) gave other reasons.

During the land acquisition processes, those whose land was either fully or partially submerged received assistance from local government officers and community leaders in negotiating cash compensation. Only those who had partially submerged land indicated that they did not receive enough cash compensation; this situation rendered such respondents landless. According to their answers, 72 respondents (33 percent) indicated that after receiving compensation cash they had almost enough to set up their new house, and 64 respondents (29 percent) indicated that they had enough for a daily meal. Only a few of the respondents indicated that their cash compensation contributed to their savings. Hence, the results seem to suggest that cash compensation alone was not enough for the resettled individuals (particularly for those whose land was only partially submerged), and they were unable to reconstruct their lives.

To date, almost all of the respondents were satisfied with the education facilities provided in the resettlement program. Also, in contrast to the first generation, the second generation suggested that their income was more stable in their present setting. This means that the first generation, especially those who live downstream, continued to struggle to adapt to this new area. They also experienced difficulty in finding a job. Most of the respondents indicated that their current situation posed greater difficulty in obtaining suitable jobs. This is reasonable, since farming was their principal former job and most current jobs required greater skills and younger staff. Despite the first-generation difficulties in adaptation, the second generation easily adapted to the new area. Regarding their housing condition, all of the residents indicated that their present home was better than their previous one.

The majority of respondents indicate that they are satisfied with their living conditions. Their houses are larger than previously and the environment is much better. The respondents who elected to live in the downstream area have better education facilities as well as more opportunities for their children to obtain higher education. Also, they have better and easier access roads to all public facilities (e.g., hospitals, banks, the city center, and others). In contrast, respondents living in the upstream area were isolated. Nevertheless, they earned more stable incomes, since they had more opportunities to work in the agricultural sectors. In general, both were satisfied with their advantages and disadvantages. Regarding community adaptation, the respondents experienced no major difficulty living in the resettled area; they were easily adapting to their new neighborhoods and environment.

Concerning the compensation processes and negotiations, the majority of respondents felt that they were fair enough, despite the fact that their expectations were not completely fulfilled. Most of the resettled individuals spent their cash compensation to purchase new land and establish a new home. Since they had moved to the remote area they could get more land compared to what they previously owned. In addition, they also owned more valuable belongings, except those who moved to the downstream area. Residents living in the downstream area had a higher economic burden than those living upstream, and as a consequence they were forced to spend much of their cash compensation.

Saguling Dam (Indonesia)

In West Java Province, the Saguling Dam was constructed during 1983–1985 with the purpose of provision of hydroelectric power to heavily populated Java and Bali. The project has displaced 3,038 families from the inundated area, and affected 7,626 families that lived outside the inundated area but had land and sources of income in it (PLN 1989). It has been acknowledged by researchers that many of the most challenging socio-economic impacts of dam construction relate to the migration and resettlement of people near the dam site or in the catchment area (Bartolome et al. 2000; Cernea 2003a; Egre and Senecal 2003).

The government provided a resettlement program, and several schemes were offered: (1) transmigration to an island outside of Java; (2) local transmigration (migration within West Java); or (3) a decision by the resettlers themselves as to where they would move. Other than that, a few additional alternatives such as estate work, construction, and agri-aquaculture were also provided as options by the government. Only 3.9 percent of the resettlers followed the first two schemes, and the rest chose to relocate near the reservoir (Suwartapradja et al. 1985). Also, some of the resettlers who chose to transmigrate to outside of Java Island and some of those who had moved out based on their own choice returned to the area surrounding the lake for various reasons. Consequently, the population density around the lake increased.

In such a case, a local resettlement scheme would perhaps assist in meeting the goal of large-scale resettlement. As a result of the environmental impact and assessment recommendations on resettlement (IOE 1979; Soemarwoto 1990), the State Electricity Company (Perusahaan Listrik Negara—PLN) implemented a local resettlement scheme. Local resettlement was to be supplemented by livelihood-rebuilding opportunities in aquaculture, tourism, and small-scale industry using the electricity provided by the Saguling Dam. The "aquaculture resettlement option" was a priority of the efforts of the dam developers (PLN) and the financier (World Bank and Japanese **Overseas Economic Cooperation Fund (OECF)**) (Soemarwoto 1990; Sutandar et al. 1990), and it would play the most important role in helping resettlers rebuild their livelihoods shortly after the inundation.

Development of cage aquaculture and capture fisheries was then initiated to provide rural jobs and maximize all possible productive uses of the new water resources. The use of floating net cage (FNC) aquaculture was proposed, because it was deemed compatible with engineering forecasts of reservoir operations and draw-downs, and also because there were many deep, sheltered bays suitable for cage aquaculture (Gunawan 1992; Nakayama 1998; Manatunge et al. 2001). In total, 1,500 families were targeted by a cage aquaculture training program conducted by the Institute of Ecology (IOE) and International Center for Living Aquatic Resources Management (ICLARM), in cooperation with the West Java Fisheries Agency, and under direction of PLN (IOE-UNPAD and ICLARM 1989).

It has been reported that the resettlement schemes for the Saguling Reservoir were highly successful, as a result of the economic and social benefits of fisheries development. This reported success is perhaps true because integrated fisheries systems have increased the number and diversity of jobs, including the number of jobs with higher pay. However, several researchers have raised serious questions regarding the extent to which the full benefits of aquaculture development are enjoyed by resettlers (Gunawan 1992; Nakayama 1998; Nakayama et al. 1999a; Nakayama et al. 1999b; Manatunge et al. 2001; Miyata and Manatunge 2004). They point out that there is a clear need to carefully examine whether resettlers have secured a satisfactory level of ownership of aquaculture development, which could be a better alternative to indemnify resettlers for their lost livelihood. It has also been suggested that non-resettlers and third parties (such as entrepreneurs) enjoy most of the benefits of aquaculture, stemming largely from lack of ownership among resettlers. That is, the capital needed to invest in and operate a few fish cages to secure an income sufficient for a single family was beyond the reach of ordinary resettlers, and they were thus obliged to work as employees, rather than owners. Shortly after inundation of the reservoirs, more than 60 percent of the resettlers remained below the Indonesian poverty line. Thus, the majority of them lacked the capital to invest in aquaculture. This situation was anticipated from the outset of the project, so the development of cage systems using cheap and affordable technology was undertaken. However, this alternative technology was not widely adopted by resettlers. One of the major reasons why resettlers failed to enjoy ownership of aquaculture development in the Saguling Reservoir was their non-acceptance of this alternative technology (Manatunge et al. 2001).

The extension program in Saguling preferentially promoted large-scale, capital-intensive systems, which required high levels of input and aimed at maximizing production. The research and training did not encourage aquaculture initiatives appropriate to resource-poor farmers, so they failed to significantly impact livelihoods and income generation amongst the poor. The project did not prioritize the opportunity to conduct research in partnership with poor farmers and to focus on technologies that support their needs. The final results of the project over-emphasized success in terms of fish production, but not the benefits to the local community or the resettlers.

The benefits of FNC aquaculture, which were originally guaranteed to the resettlers through provincial legislation designed to give them exclusive ownership over both production and marketing sectors of the industry, were usurped by politically powerful non-settlers and consolidated in the hands of the urban rich from Bandung and Jakarta (Gunawan 1992). Most resettlers are involved principally as employees of absentee owners. It is difficult to discern if the evident increase in economic well-being of the resettlers is due to these new, high-paying aquaculture jobs or simply due to the increased economic opportunities that were available in most of Java in the 1990s.

The dynamic nature of social systems is perhaps the least understood factor of influence during and after dam construction. For instance, outsiders may seek to take advantage of newly created commercial and social opportunities by moving into the area. The "middlemen problem" is one of the most serious socioeconomic problems in aquaculture development (Manatunge et al. 2001), which eventually leads to local fishermen's earnings being reduced to unfair and unacceptable levels (Gunawan 1992). The case of Saguling demonstrates this quite clearly. The lack of fishery regulations made it difficult to control the influx of influential outsiders or to limit their activities, which hindered the involvement of the poor. Laws concerning fishing rights and issuance of "proper permits" need to be formulated, and enacted laws need to address the roles of various groups to ensure that the practices of middlemen are not detrimental to those affected by reservoir inundation, and to maintain equity among the latter. Only under a clear legal mandate, with strong community backing, can the prospects of a socially stable and sustainable industry to restore resettlers' livelihoods be realized.

The initial environmental impact assessment proposed that the aquaculture capacity of the reservoir was 5,800 cages. It was expected that they would be dispersed throughout the most suitable sites (deep, sheltered, and well-flushed bays), but this distribution did not occur. Farmers crowded their cages into a small number of bays due to better availability of economic infrastructure and better access to large fish markets. This crowding led to pollution due to waste feed and nitrogen discharges from the cages (and the waste of an increasing number of residents at the water surface). The result has been the development of nuisance algal blooms and, more frequently, oxygen depletion, leading to large fish kills (Costa-Pierce 1997; Djuangsih et al. 1997). The water quality of the reservoir has degraded rapidly over the past decade. The condition of the reservoir has been categorized as hyper-eutrophic (Djuangsih et al. 1997; Costa-Pierce 1998). Non-point source pollution has also contributed considerable quantities of nutrients such as nitrogen and phosphorous from surface run-off. The increased erosion in riparian areas has contributed to higher rates of siltation, and it is suspected that approximately 40 million tons of sediment has already silted at least a quarter of the reservoir capacity.

As previously mentioned, farmers tended to crowd into a few selected bays due to better facilities, which led to high levels of pollution and massive fish kills. The haphazard and unsustainable development of aquaculture has threatened future prospects for fisheries in the reservoir. Aquaculture should have been developed in a more eco-friendly and less intensive manner, with less polluting methods and the introduction of no-feed aquaculture. High organic wastes in the water column have caused low concentrations of oxygen in most parts of the reservoir. The fluctuation in dissolved oxygen to levels below lethal limits for fish life has caused spectacular fish kills in recent years (Djuangsih et al. 1997). Fish are fed with excessive amounts of feed to stimulate growth, and it has been estimated that at least 3 percent of the feed introduced into the water is not consumed by fish and is added instead to the water column (Costa-Pierce 1997).

The suspended-solid content of the water column is extremely high, and could reach higher levels with re-suspension of bottom sediments caused by high wind conditions. The turbidity of water is highly variable, with values exceeding 400 nephelometric turbidity units at times. These conditions have persisted for most recent years, drastically decreasing fish production. Iron and ammonium concentrations, together with several heavy metals, have been detected at levels too high to allow healthy fisheries activity.

Due to over-exploitation of aquaculture resources, the potential for increasing yields in the long run is extremely limited. Water is the most important input for fisheries and a key element in the success of these projects. Poor water quality may impair the development and growth of fish. It may also degrade the quality of the product by tainting its flavor or by causing accumulation of high concentrations of toxic substances that endanger human health (Costa-Pierce 1997).

A survey of the resettlers

In 2011, a survey was conducted at two inundated villages: Bongas and Sarinagen. The former lies in the southern sector of the Saguling Dam, which in general has better water quality compared to other parts of the reservoir, such as Sarinagen in the northern sector. Bongas was selected as the site of cage aquaculture program implementation and the center of training and supporting activities. It is proposed that the success of the cage aquaculture program in Bongas was due to better water quality and socio-economic and infrastructural support. When we conducted the survey in 2011, the practice of FNC aquaculture in Bongas was still common, while it was low in Sarinagen.

The household heads interviewed were those who were at least 17 years old (or married) at the time of dam construction. It was assumed that they had strong perceptions about any impacts of the dam construction in the past and can clearly provide any relevant information. The total number of respondents was 147 resettlers, consisting of 97 persons from Bongas and 50 from Sarinagen. As the population of the resettlers was unknown, non-probability sampling was employed. Targeted respondents were traced through the list provided by the village administrative offices.

With regard to their occupation, primary attention was on any changes of employment caused by dam development. As people were involuntarily resettled from their original settlements to unplanned sites, it resulted in some socioeconomic hardships, including changes in employment and income-generating opportunities. The number of self-employed farmers decreased drastically after resettlement, both in Bongas and Sarinagen; in contrast, the number of share-croppers increased. In addition, the survey found that in the present time the number of unemployed resettlers has increased markedly, both in Bongas and Sarinagen. Certainly, many reset-tlers who were interviewed are not young anymore; however, in rural areas it is common for old men to continue to work for a living. At present, the number of people employed in public offices, private sectors, and as labor-ers is stable. Meanwhile, other sectors such as business, entrepreneurs, and services employ a considerable proportion among the resettlers (i.e., from 23 percent to 30 percent). Tourism and estate sectors developed by the local government and PLN were not successful (see Table 2.2).

Table 2.2 Types of resettlers' occupations before resettlement and at present in the villages of Bongas and Sarinagen

Occupation	Category	Bongas		Sarinagen	
		Before resettlement	*Present*	*Before resettlement*	*Present*
1 Farmer	Self-employed	74 (76.3%)	11 (11.3%)	47 (94%)	14 (28%)
	Share-cropper	7 (7.2%)	43 (44.3%)	2 (4%)	26 (52%)
2 Employee	Public offices	6 (6.2%)	8 (8.2%)	6 (12%)	2 (4%)
	Private sector	3 (3.1%)	1 (1.0%)	0 (0%)	0 (0%)
3 Laborer	Farm	5 (5.2%)	4 (4.1%)	0 (0%)	2 (4%)
	Unskilled manual	4 (4.1%)	1 (1.0%)	1 (2%)	0 (0%)
	Construction	1 (1.0%)	2 (2.1%)	2 (4%)	2 (4%)
4 Unemployed		1 (1.0%)	23 (23.7%)	0 (0%)	7 (14%)
5 Others		27 (27.8%)	23 (23.7%)	4 (8%)	15 (30%)

Note: Interviewees were allowed to give more than one answer to this question. Survey conducted in 2011.

Source: Authors.

At the beginning of the aquaculture program, implementation of FNC aquaculture development was aimed at absorbing persons unemployed due to dam construction, where the Bongas area became the central site and has remained the predominant center for FNC aquaculture. A few years after its development, however, the FNC population at the Saguling Dam declined continually due to fish diseases, poor water quality, and as a consequence of the national economic crisis in Indonesia in 1998. Most of the resettlers in both villages were not able to reconstruct their FNC aquaculture due to limited access to capital and assistance. Our survey showed that the resettlers engaged in FNC aquaculture in Sarinagen are much fewer in number compared to those in Bongas (Table 2.3), although resettlers in Bongas told us that their cage possession was less than expected. Interestingly, some interviewed resettlers explained that they have only a small part of the FNC population; most are owned by non-resettlers who have capital and generally come from outside Saguling. The data of FNC ownership from the two villages also indicated that there was an inequality of access among the resettlers to the designed cage aquaculture program.

It was revealed from the interviews that the resettlers, both in Bongas and Sarinagen, enjoyed the new environment with increased quality of livelihoods. Changes in house size and style, connection to the electricity grid, availability of drinking water, health facilities, and education facilities are among the improved measures compared to their living conditions before resettlement. However, resettlement programs set up by PLN and the local government do not seem to have fulfilled expectations completely, with employment being one of their main concerns. Among the resettlers who live in Bongas, there is a decrease in satisfaction level with the present jobs, while among those living in Sarinagen there is a slight increase. The data indicate (see Table 2.2) that more resettlers lost their original jobs as self-employed farmers and became share-croppers. This might be due to a larger number of unemployed resettlers in Bongas as well. The type of employment in which the resettlers participate will affect the level and stability of income, and, hence, economic conditions. Therefore, the feeling of dissatisfaction with the present jobs in Bongas may correlate to their dissatisfaction

Table 2.3 FNC aquaculture ownership

Ownership	Bongas	Sarinagen
1 Households own FNC		
1985–1987	55	18
2011	11	4
2 No. of cages/household		
1985–1987	15	7
2011	7	8

Source: Authors.

with their present economic conditions. In contrast to what happened in Bongas, the majority of resettlers from Sarinagen (56 percent) expressed their satisfaction with their present economic conditions.

With regard to the present living environment, the majority of resettlers who live in Bongas (64 percent) and in Sarinagen (60 percent) expressed satisfaction. They preferred living in the vicinity of the Saguling Dam rather than to accept the resettlement schemes provided by PLN and the local government. The new settlement may not be a cause of stress for them, as they share the same culture and customs. However, the majority of resettlers do not seem to have access to public facilities such as community halls, public toilets, volleyball and badminton courts, and football fields. For rural areas, local governments appear to give a lower priority to the development of such public facilities, compared to other infrastructure projects, such as roads, bridges, and irrigation facilities.

The majority of resettlers who live in either Bongas or Sarinagen feel happy with respect to the present environment, which promotes the interests of their children. They believe their children have better opportunities to get a good education, so that in the future they can be better off than in the past. It is also surprising that even though land ownership has decreased, they are satisfied with the land availability now. This could be a sociocultural case: typically, people in West Java like to stay close to their relatives rather than being separated. As long as they can mix with their family members, ownership of property becomes a lesser priority. For them, the level of job opportunities for their children is satisfactory at the moment. They are optimistic that their children will have better livelihoods compared to their parents. The resettlers hope that, with better education, their children will have employment opportunities at public offices, schools, military services, private companies, and other sectors.

Conclusions

The survey revealed that the construction of the Saguling Dam has had long-term socioeconomic impacts for resettlers. This is true because, about 25 years after relocation, the resettlers still suffer from unintended occupational changes. Many resettlers involuntarily converted their occupations from self-employed farming to share-cropping, which has lower benefits in terms of income-generating opportunities and social status. After losing their fertile farmland, they were not able to purchase the same amount of land in the new settlement with compensation money received because land prices increased as a result of the dam project. Another important finding is that many resettlers remain unemployed, which indicates fewer job opportunities since resettlement. Both these negative impacts seem to be common consequences resulting from involuntary resettlement programs induced by dam construction projects.

The rapid development of FNC aquaculture did not mean that this farming activity was without constraints. Catastrophic fish mortality became one of the major factors affecting the sustainability of aquaculture, and the unfavorable environment has made incidents of high fish mortality frequent.

The present study reveals that during the 1990s many resettlers who owned FNCs were not able to reconstruct the aquaculture operations after incidents of fish kill, and FNC aquaculture is capital intensive. The resettlers had to expend about IDR 2.4 million (USD 270) in cages and operation costs per household (Gunawan 1992). In this situation, the capital needed to invest in and operate a few fish cages to secure an income sufficient for a single family was beyond the reach of ordinary resettlers, and they were obliged to work as employees rather than owners. After the Asian financial crisis in 1997, it was even more difficult for them to reconstruct their FNC aquaculture. Some interviewees stated that non-resettlers and entrepreneurs had been dominating the ownership of FNCs in the Bongas area right from the beginning. The researchers consider that in the long term, aquaculture development has failed to offer alternatives to indemnify resettlers for their lost livelihoods.

However, in general, the livelihoods of the resettlers did improve in their new settlement. They have been enjoying infrastructure facilities and services that were not available in their original settlement. Nevertheless, some people were not satisfied with their present jobs and economic conditions, which can be considered to be the most important aspect in their interest.

Kusaki Dam (Japan)

Negotiations between resettlers and developers

The Japanese government enacted the Multiple Purpose Land Development Act in 1950, aiming to revitalize agriculture, flood control, and industry after World War II. In Gunma Prefecture, the Comprehensive Development Plan for the Tone Specific Area River Development Plan was designed to promote agriculture and electricity production, as well as the forest industry. Under this plan, multi-purpose dam developments were planned in Gunma Prefecture from the late 1950s to the 1960s. Some 221 households were resettled as a result of the Kusaki Dam Project. Of these households, 103 (about 46 percent) moved out of the village, and 94 of these households moved to neighboring towns (Mizushigen Kaihatsu Koudan 1974). The rest resettled within the same village.

The Kusaki Dam was the first multi-purpose dam constructed on the Watarase River. After the pilot investigation for the dam construction was initiated by the Ministry of Construction in 1958, the affected communities gradually set off protests against the dam construction, as part of the village would be submerged. The village of Azuma was the only village affected by the Kusaki Dam construction. Rock quarrying and stone cutting were the main local industries before resettlements and continue to be to this day.

Among the affected communities, the Kusaki district was one of the most significantly inundated; residents of this area took the lead in establishing the Alliance for Construction Resistance with other affected districts in 1963 (Seta gun Azuma Mura 1998). The aim of the Alliance was, in practice, not to protest against construction of the dam but to secure generous compensation packages. Those involved in the Alliance were stone cutters, since the majority of resettled individuals were involved in the stone industry; the collective actions of these stone cutters resulted in coherent requests.

At the same time, the Azuma village office launched a committee for dam construction, which served as a focal point for the Water Resources Development Public Corporation (formerly, the Japan Water Agency; hereafter, WRDPC). The Village Assembly also established a special committee for dam construction, whose aim was to study the issues related to dam construction and to implement provisions. The prefectural government and the Assembly initiated mediation talks, hoping for a breakthrough in the deadlock between the WRDPC and the villagers. Petitions from the Alliance were delivered to the prefectural government and the WRDPC via the village committees. According to the petition, dated July 1966, the committee requested compensation for livelihood rehabilitation, including alternative land for homes, public compensation, promotion of local industries, and tourist development (Seta gun Azuma Mura 1998).

In response to these petitions several joint field missions by the prefectural government were dispatched to Azuma. These structural negotiation units attempted to identify and consolidate villagers' requests. After many meetings between the village and the WRDPC, the villagers changed their attitude from "completely opposed" to "opposed with conditions" in 1965. An investigation of land boundaries began in 1968. Prior to this investigation, the village asked the WRDPC to announce provisional compensation rules and rehabilitation plans for the stone industry. The village also asked the WRDPC to set a minimum level of compensation for those in the village (Seta gun Azuma Mura 1998).

The villagers also submitted a petition for livelihood rehabilitation, since construction related to the dam had already begun; villagers were concerned about the lack of provisions for livelihood rehabilitation, despite the fact that the project had already started. The Council for the Countermeasure of Rehabilitation of Livelihood was established in 1968 to discuss resettlement measures. Throughout this consultation, the WRDPC took the stance that negotiations for both general and public compensation would take place parallel to one another; however, the village office was strongly against such parallel negotiations, since they believed that once the construction began their negotiation for public compensation would have been less productive. These actions demonstrate the high level of concern the villagers had about compensation for livelihood rehabilitation. After the Alliance and the WRDPC reached an agreement (that included an explanatory meeting for the compensation offered) the Kusaki Dam project moved forward. The WRDPC began construction of the Kusaki Dam in 1971 after the conditions of the public and general compensations were finally agreed upon.

Compensation schemes

Several compensation measures were carried out in order to facilitate the resettlement of those affected by the Kusaki Dam project. The Alliance for Construction Resistance played a key role in the negotiation of these measures. The Alliance had significant input into the pre-construction aspects of the project, including field visits to existing dams and provisions for rehabilitation. Throughout the process of discussions the Alliance gradually identified its requests and set its target for compensation. According to an interview with a son of a core member of the Alliance, "[t]he ultimate purpose of the alliance was to gain compensation for houses, even for people that did not own a house and leased a plot at the time" (interview by authors, 2011). Also, the petition submitted to the WRDPC in 1968 inquired about the minimum level of compensation that would be given (Seta gun Azuma Mura 1998). The Alliance faced many difficulties throughout negotiations but gradually achieved its objective of obtaining a minimum level of compensation for its most vulnerable members.

An evaluation of material loss entails counting the items lost in exact detail, from trees to houses; this method creates the incentive for residents to maximize their count of losses. The resettlers interviewed by the authors were under the impression that they could receive cash compensation for their material loss in accordance with the cash compensation rule. In fact, some resettlers planted additional trees just prior to the evaluation so that they could receive more compensation than others around them. As a matter of fact, the appraisal of material loss was often increased by a nominal amount even after the national Guidelines on Standards for Compensation for Losses Associated with the Acquisition of Land for Public Purposes (see Box 2.1) was enacted in 1962 (Maruyama 1986). The resettlers were focused on receiving their full compensation and rarely cared about the details of what they were being compensated for.

The national forest adjacent to the village of Azuma was handed over to the stone cutters as a source of more stone. Financial support for drivers' licenses was also provided by the WRDPC to improve mobility and alternative transportation. In this way, some resettlers could begin their own businesses as carriers in the transport industry. The bus service helped children and resettlers near the dam sites to commute to schools and workplaces. Those interviewed indicated that these compensation measures were effective for securing jobs, convenience, education, a future for their children, and mental stability for those who stayed near the dam sites.

Lessons learned

Negotiations for public compensation throughout the Kusaki Dam project was settled before negotiations for individual compensation began. The public compensation allowed individuals much freedom to relocate near the dam site. The stakeholders collaborated to obtain comprehensive compensation deals from the WRDPC compensation schemes, and there was a measure of coherence in the villagers' thinking as well as having their leader present for negotiations. The Kusaki project demonstrates that it is possible to generate acceptable compensation, and therefore it is important to allow time to discuss the real needs that are likely to arise after resettlement and to communicate these needs to the dam developers. Also, the structural negotiation units in this village successfully supported the progress of step-by-step negotiations. Consequently, the active involvement of villagers in consultation meetings allowed them the opportunity to discuss the future of their village.

The measures taken for livelihood rehabilitation are key factors that allow resettlers to remain living near a dam site. Without a vision for future development, the resettlers find it difficult to determine whether or not they want to remain nearby.

The resettlers formation of the Alliance for Construction Resistance was an effort to consolidate their requests and to negotiate with the dam developers. The items discussed with the developers included rehabilitation of livelihood, construction of collective houses and bridges, development of an alternative quarry, and promotion of tourism. The Alliance also considered measures for individual compensation, including houses for the most vulnerable households. Based on the interviews, the measures put in place for the vulnerable were considered as a part of the provision for depopulation.

The village of Azuma requested a detailed list of the provisions offered, in an attempt to secure the existing industries (e.g., forestry, agriculture, and the stone industry). The developers secured the villagers' mobility by providing financial support for drivers' licenses, as well as public transportation. Furthermore, almost half of the resettlers were engaged in the stone industry (Mizushigen Kaihatsu Koudan 1974); therefore, it was important to secure a quarry for stone materials. Azuma requested a land grant from the government at the beginning of their negotiations. Although the primary purposes of these measures (e.g., land grant from the government, financial support for drivers' licenses) are to assure the quality of livelihood, the case demonstrates that the measures were conducted in characteristics of public compensation. The village suggested that the provisions required to secure the existing industries were significant enough to contribute to community preservation. As a result of these consolidated requests, it held a strong bargaining position in the negotiations.

The policy for resettlement and livelihood rehabilitation, stipulated by the Act on Special Measures concerning Measures Related to Water

Resources Areas (see Box 2.1), was not applied in the Kusaki Dam project, as the planning of these dams commenced prior to 1973. An investigation into the actual state of livelihood rehabilitation for the resettled individuals was conducted by the WRDPC and related institutions after completion of the project. The investigations revealed that the resettlers felt that the provisions for livelihood rehabilitation were not fully implemented. For example, the resettled individuals were asked what kind of provisions were necessary for livelihood rehabilitation, and therefore they expected their voice to be heard and greater effort placed on job transition and resettlement (Nihon Dam Kyoukai 1978; Mizushigen Kaihatsu Koudan 1974). The measures for livelihood rehabilitation had to be implemented in order to mitigate the negative physical and mental impacts on the resettled individuals. The resettled individuals were looking for "visible compensation" (i.e., compensation by which they could realize that they had actually been compensated) as much as possible.

Despite the absence of a definitive concept of livelihood rehabilitation, other special arrangements (i.e., support for grants for the quarry and drivers' licenses) greatly improved the livelihood rehabilitation of the residents. Also, the Alliance set a minimum target for the negotiations, especially on compensation pertaining to the most vulnerable members. Without adequate compensation for livelihood rehabilitation, they would not be able to decide if they should remain near the dam site or move further away. The project demonstrates that flexible financial and non-financial compensation measures not only serve as a catalyst for community rehabilitation but also as a bond for communities. Regarding the retention of social bonds and communities, the conventional approach to compensation, such as monetary compensation for material loss, may still fail to secure the fullest extent of livelihood rehabilitation. It is important to show the best available compensation measures for livelihood rehabilitation in advance, and also to include a safety net for the vulnerable to reduce the emotional strain on the resettled individuals and to contribute significantly to community preservation.

According to Maruyama's study on the correlation of the period of compensation negotiations and the degree of difficulty (the concept of difficulty is determined by examining the overall consequences of various conditions, such as the unique characteristic of the dam and the area), the period of compensation negotiations would be influenced by the number of difficulties induced by these conditions in the process of compensation negotiations (Maruyama 1986). These increased difficulties tended to result in a longer period of negotiation. The Kusaki project shows that the presence of a definite consensus will unite villagers and allow them to consolidate their requests so that there is more clarity in the resolution. Also, structural negotiation units with strong wills can overcome many difficulties. In order to avoid an unnecessarily long negotiation process, the structural negotiation units for resettled individuals should hold inclusive stakeholder dialogs, as these will greatly bolster the confidence of the stakeholders.

The Kusaki project taught us that, as perceptions change about dam construction projects throughout the negotiation process, several positive effects can be seen in community rehabilitation. Some of following reasons why villagers agreed to the dam construction can be deduced from the interviews: (1) the villagers could negotiate in their favor and take time to prepare for the resettlement through their active participation in the negotiations; (2) the majority of resettled individuals who were engaged in the stone industry were able to continue in their jobs and purchase new heavy machinery that would be necessary to continue their work; and (3) the village of Azuma could gain a large amount of fixed property tax from the dam. The village of Azuma and its villagers gradually saw the resettlement and dam construction as a "new opportunity" for development in the area. They believed that the public compensation would contribute to regional development and that villagers would have the opportunity to live in Azuma. In fact, the village has been promoting tourist businesses after the completion of the dam. Negotiations with strong opposition parties will often lead to a deadlock; however, changes in perception can allow a reconsideration of alternative choices. Therefore, it is important to have open discussions and to learn from other dam projects throughout the process of negotiation. The affected individuals can consider their individual compensation along with the dam construction plan and decide where they will live in the short term. To garner better decisions for the resettled individuals, the affected people need to learn of the pro and con viewpoints from other experiences. Developers are typically passive when it comes to providing opportunities to learn from the experiences of others because of a perceived risk that this may complicate or negatively influence negotiations. Instead, however, developers should support and provide opportunities to create better dialog among villagers; continuous discussions between the developers and villagers would result in better solutions and indirectly contribute to community continuity.

Jintsugawa Dams (Japan)

A unique compensation scheme was once adopted for resettlers of three small-scale dams constructed in the early 1950s along Jintsugawa (the Jintsu River) in Toyama Prefecture in Japan. The land that resettlers owned (and still own) was not bought or compensated for by any alternative land. Resettlers still own their land, now underwater, for which they constantly receive rent from the owner of these dams, now for over more than half a century. Jintsugawa Dai-ichi Dam, Dai-ni Dam, and Dai-san Dam (Jintsu River Dams No. 1, No. 2, and No. 3) are located within a radius of just a few kilometers, and the dams were built by the district electric power company, Hokuriku Denryoku.

These Jintsugawa Dams were constructed at a time when Japan was about to take off with rapid economic growth. In the early 1950s, Hokuriku Denryoku was established following a major restructuring of the power industry by the government. Hokuriku Denryoku was in urgent need to increase its power generation capacity in order to meet the rapidly increasing demand from the growing manufacturing industries in the area, such as aluminum factories. Three Jintsugawa Dams were planned and constructed in a great hurry. Intensive negotiations with the local communities living in the dam sites over conditions for compensation and resettlement took place and were concluded in a short period of time. In the course of a few years, the proposal was discussed and approved by the local communities, followed by construction of the dams and commencement of power generation (see Table 2.4).

The reason why Hokuriku Denryoku needed to develop hydropower stations in such urgency was because the existing large and efficient hydropower stations in the region (Hokuriku Region) were attached to another regional power company, Kansai Denryoku, which had been serving the already-industrialized area in Osaka and Kobe regions in the western part of Japan. Hokuriku Denryoku was thus obliged to build hydropower stations to meet local demand (Hokuriku Denryoku 1962).

Rent scheme proposed and adopted

In the process of negotiations with the power company, landowners (both resettlers and non-resettlers) refused to sell their land, but instead suggested that the "rent" scheme be adopted so that they would maintain their ownership of the land. Hokuriku Denryoku initially proposed to buy up all the land of the dam sites. However, at that time the company had realized that it did not have enough funds to pay for all the land all at once. So, pressed by the urgent need of constructing these dams, the company agreed on the scheme. While agreements with some different conditions were drawn up

Table 2.4 Chronology of negotiations and construction of Jintsugawa Dams Nos. 1, 2, and 3

	No. 1	*No. 2*	*No. 3*
Negotiation started	1951	1952	1953
Construction base set up	1952	1952	1954
Agreement with local communities	1952	1952	
Construction started	1952	1952	Right bank: 1954 Left bank: 1955
Construction completed	1954	1953	1954
Operation started	1954	1954	Right bank: 1955 Left bank: 1956

Source: Japan Dam Foundation (2008) and Hosoiri-mura (1987, p. 626).

between Hokuriku Denryoku and different settlements, in all the agreements regarding Jintsugawa Dams No. 1, No. 2, and No. 3 the same type of rent scheme was adopted.

The rent schemes adopted for the three Jintsugawa Dams as a means of compensation were the only deals of their kind that ever materialized in Japan, and they have survived to date. As is often the case with land submersion related to dam construction, the local communities of these dam sites were old farming villages engaged in rice and vegetable cultivation, sericulture, and so forth. When construction of the dams was proposed, many in the local households were part-time farmers, working on their own or on somebody else's farm while having some other occupation to make a living. Having lived on and earned income from the land inherited from their forefathers, the local people had a strong sense of reliance and attachment to it, not only in the practical but also the emotional sense (Nakayama and Furuyashiki 2008). Details of submerged land area are shown in Table 2.5.

The "rent" for the submerged land was initially calculated in the following manner:

1   For paddy fields, the "governmental purchase" price and the average yield became the basis of the "rent." (At that time, the government purchased all rice crops.)
2   For dry farmland, 70 percent of the "rent" of paddy fields.
3   For forest, 10 percent to 15 percent of the "rent" of paddy fields.

In addition, as a show of gratitude to the resettlers, 302,510 yen and 211,757 yen were paid per hectare of paddy field and dry farmland, respectively. For residential plots, 7,654 yen was paid per square meter, and new

Table 2.5 Area of submerged land and the number of resettlers compensated for the construction of Jintsugawa Dams Nos. 1 and 2

	No. 1	No. 2
Paddy field (ha)	5.9	8.4
Dry farmland (ha)	11.3	0.6
Residential plot (ha)	0.96	0.5
Forest (ha)	25.1	6.3
Grassland (ha)	2.3	–
Other (miscellaneous) land (ha)	0.5	1.5
No. of relocated households	26	21
No. of relocated religious buildings	1	–
No. of relocated communal buildings	–	1

Note: For Jintsugawa Dam No. 3, only two families had to be relocated (Hokuriku Denryoku 1962, p. 172).

Source: Souri-fu Shigen Chousa-kai (1954, p. 295).

houses were given to the resettlers free of charge (Miyashita 1961). Some communities also had other conditions agreed for communal benefit, such as construction of a new communal swimming pool, as the village children would no longer be able to swim in the river.

While the rent was initially determined according to the government purchase price of brown rice, the rent was later (in the 1980s) switched to fixed-rate payment. Up until some 20 years ago, the government's purchase price of rice had been kept high enough, profiting the landowners favorably when they received the rent according to the rice price. However, once the government's rice price started to stagnate such merit disappeared, and Hokuriku Denryoku and the villagers agreed that the rent should be decided independently from the official rice price.

Since then, the amount of the rent has been revised every ten years. The revised rent is first negotiated and informally decided between Hokuriku Denryoku and representatives of the relevant communities, and then company staff visit every recipient household to obtain consent. Some may express dissatisfaction over the revised rate, but different rates cannot be applied to different households. Some of the landowners now live away from the village (outside the prefecture), and with them the company simply negotiates (gets consent on the rent) over the phone. According to Hokuriku Denryoku, the rent paid to the villagers and communities is currently set somewhat higher than the market price of real estate for paddy fields, dry farmlands, and forests in the area.

Perception of resettlers

Several settlements were affected by the construction of the three Jintsugawa Dams, and a total of 49 households were obliged to relocate. Most of them stayed in the same area, having new houses built on the upper part of the hillside by the power company.

The authors interviewed some resettlers in February and December 2004. All those interviewed expressed their satisfaction with the rent scheme. They were proud of the fact that this scheme has been implemented for many decades, even if the amount of the rent may not be high enough to greatly enrich their daily life today. They were pleased that they had kept ownership of the land. Also, by renting out their own land to Hokuriku Denryoku and receiving rent money every year from them, psychologically, the local villagers seem to have been able to keep their position higher than or at least on an equal level with the power company.

Resettlers also find the rent scheme is important for them to guarantee a continuous food supply. The villagers stated that if the land had been sold off for a large sum of money once and for all, such money would not have remained in their hands for long, but would have been used up in a short period of time. In fact, a few years after the rent scheme had started some recipient families opted to sell their land to the power company and purchased securities, only to see a great loss afterwards. Having seen such cases, the interviewed villagers were convinced that it was far better to continue receiving a certain amount of money every year. Also, losing the land forever signified losing their ancestral profession, since farming consisted of a significant part of the villagers' means of life in those days. Therefore, rather than losing their land and profession at the same time, it was felt that it was better to rent out the land and seek other means of income with the secure knowledge that there would be a certain amount of income in the form of "rent" every year. Some villagers also commented that the rent scheme has served well in the sense that it has benefited not only individuals but also the settlement as a whole (Nakayama and Furuyashiki 2008).

Therefore, the rent scheme employed in this case might have helped alleviate the sense of victimization by the dam construction, mentally as well as practically. The authors believe that these resettlement compensation schemes are innovative, and that such schemes may be applied in developing countries about to take off in their development. The application of such schemes might have reduced many of the problems of some projects implemented in the developing world.

Lessons learned

The following lessons were learned through a literature survey, field visits, and interviews with relevant people. The rent scheme was applied in Japan only for the Jintsugawa Dams, for the government of Japan in 1953 established the principle for compensation against lost lands, etc., due to inundation by reservoirs for the sake of hydropower generation (Government of Japan 1953). Similar rent schemes may not be applied in Japan in future, as long as the existing principle is valid. The rent scheme as well as the following lessons may be applicable and useful for the developing world, if a rent scheme is being considered and implemented in future.

Sustainability of the rent scheme

The rent scheme proposed and implemented for the Jintsugawa Dams is in its nature sustainable. Resettlers maintain ownership of the land they owned before construction of these dams, and they receive rent as long as they possess the land, even if it is now underwater. As revealed through interviews with resettlers and their descendants (who receive the rent), this measure mitigates the resettlers' feeling that they were victimized by the dam and saves them from having to sell their ancestral lands. Also, resettlers do not receive a big chunk of money all at once, which effectively prevents resettlers from wasting their compensation given in monetary form.

Resettlers and owner of dams should have an agreement on the ways and means of determining the amount of rent. Persuasive approaches include calculating the rent in accordance with: (1) rent of residential plots in the land market; and (2) the agricultural productivity of farmland in the region. It is important to carry out an in-depth baseline survey on the productivity of land in the region, and the survey should be carried out with the participation of resettlers, so that all parties concerned agree on a single way of calculation of the rent for each piece of farmland.

Risk of moral hazard

The staff members of Hokuriku Denryoku half-jokingly told the authors that with rent they have paid some two to three times more money to the owners of the submerged lands than it would have cost to buy all the lands at the beginning of the project. This seemingly higher cost of the "rent scheme" is, in a sense, rational, for the rent scheme effectively transforms the former landowners into "shareholders" in the new dam (Cernea 2003b). These "shareholders" are by definition entitled to a fraction of the revenue from the dam for their contributions of land. The rent scheme, however, might cause moral hazards among the resettlers, because they will never face any risk in their receiving rent from Hokuriku Denryoku.

In the cases of the Jintsugawa Dams, luckily, few moral hazards have been reported, presumably because the rent from Hokuriku Denryoku represents a small portion of income for the resettlers. They used to have a small area of land, mostly only to cultivate rice and vegetables for domestic consumption, and farming was not their most important source of income. They all had other occupations (e.g., as staff members of local government, workers in local mines, etc.) as their major sources of income.

The authors feel that the moral hazard was avoided in this case due to this limited importance of rent as revenue for resettlers. If a rent scheme is to be implemented for a number of households in a future dam construction project, more consideration may be needed of ways to avoid moral hazard, because creating numerous "pensioners," in particular among the young generations, may lead to many societal conflicts between resettlers and non-resettlers.

Duration of rent scheme implementation

Maintaining a rent scheme is an everlasting enterprise for the owner of dams. Hokuriku Denryoku spends a lot of time and effort in the negotiation of rent revisions every ten years. It has turned out that it would have been more cost-effective for Hokuriku Denryoku to buy up all the land at the initial stage rather than continuing to pay rent indefinitely. From an economic point of view, Hokuriku Denryoku would rather stop the rent scheme, even now, by purchasing all the land currently being rented. Nevertheless, after paying rent for over half a century, such a proposal does not seem feasible from the company's perspective, and the question would arise as to why they had not purchased the land earlier when funds became available.

Therefore, unless the suggestion of selling off the land comes voluntarily from the landowners (including resettlers) themselves, Hokuriku Denryoku must continue paying rent every year. While the company has always been ready to buy up lands from landowners (if they wish to sell), it seems unlikely that such a day would come when all the landowners give up their ownership, since, as the company officials observe, it appears that in the course of the past 50 years the annual rent payment has become an established part of the villagers' lives.

It may alleviate problems from this perpetual operation if some agreement could be made between the resettlers and the owner of a dam regarding the duration of the rent scheme, namely, whether the scheme continues in perpetuity or just for a certain number of years.

Procedures for reviewing the scheme itself after implementation for a certain number of years—to determine if the scheme would continue or not—need to be discussed before implementing the scheme between resettlers and the owner of a dam. Resettlement cases in Japan in the past show that often there exist differences in opinion among generations in the same family on whether or not land should be sold for the sake of dam construction. The older generation tends to want to keep the land, while the younger generation is more flexible on selling it.

The revision or cancellation of the rent contract, including whether the scheme should continue or not, therefore ought to be discussed with resettlers as landowners when ownership of a land changes. For example, the death of a family member belonging to the older generation may provide a resettled family with an opportunity to reconsider both the advantages and disadvantages of the rent scheme for them. Once the livelihoods of resettlers have been rehabilitated for a number of years in the new place, the owner of the dam might wish to proactively initiate discussions with resettlers about the future of the rent scheme, to mitigate problems associated with an agreement in perpetuity.

Discussion

It can be concluded that the rehabilitation of resettlers' lives in five cases out of the six referred to in this chapter was generally successful, while there have been problems in the case of Saguling. Nonetheless, the resettlers more or less expressed satisfaction with their present lives, although some issues should be noted. First, more than a few decades (half a century in Ikawa and NN1) have already passed, and much of the improvement of their lives—in particular, improvements and increases in resettlers' wealth— cannot be attributed to the resettlement programs, but, rather, to the economic development of the countries. Moreover, some resettlers seem to avoid directly expressing their satisfaction. For example, a majority of resettlers of NN1 replied "Don't know" to questions inquiring about their level of satisfaction.

Resettlement of NN1 presents an exceptional case. No resettlement program was planned, and resettlers did not participate in the project preparation or implementation, since the project was implemented during the civil war, and the necessity of participation was not recognized. Differences in living conditions and incomes of the two resettlement villages are obvious, and these can be attributed to the differences of infrastructure. On the one hand, Pakcheng is connected to urban areas by road, and has an irrigation system provided by an agriculture project between 1980 and 1995. On the other hand, Phonhang is not connected to urban areas and has only a poor irrigation system. These cases demonstrate the effectiveness of infrastructure provided, even without any participation of resettlers.

Road connection has played an important role, not only in NN1 but also in other cases. In Ikawa, the improvement of the road connection significantly impacted not only material life but also people's mind-set, and the resettlers wanted further road improvements. Villagers living in upstream resettlement areas of the Wonorejo Dam have become more isolated than those downstream due to the poor road connection. Provision of a new road would likely improve their incomes and income stability by improving better access to markets.

In the other five cases, resettlement programs were planned, and resettlers more or less participated in the program formulation, with at least their intentions surveyed. One of the important factors for project planners is paying adequate attention to resettlers' choices. If planners pay adequate attention to resettlers' potential choices and strategies in coping with the forced relocation, they may be able to devise better resettlement options. In doing so, it is also necessary to keep in mind the fact that resettlers do not perceive the options presented in the same way as the planners. For example, an optimistic resettlement plan to increase or restore income in the new resettlement area with new agricultural production or technology may not be so attractive to the resettlers. They may consider the option to be risky, as shown in Ikawa, or they may have much stronger intentions to continue to stay near their old domiciles, as shown in Wonorejo and Saguling. In this regard, the participation of resettlers in the planning of a resettlement program should not be treated merely as an occasion at which planners can persuade resettlers or make themselves understood. On the contrary, it may be a valuable opportunity for planners to learn about resettlers' possible choices and strategies. Intensive negotiations in the case of Kusaki led to successful planning of the resettlement program. If resettlers feel and are convinced that they have consciously chosen the best option among the possible choices, even though there are some constraints at the time of the decision, they may find their choice satisfactory in the long run. As mentioned earlier, resettlers should not be treated as mere beneficiaries or victims of resettlement.

Through surveys in Wonorejo and Saguling, the planners knew that few resettlers wished to participate in the TP, contrary to their initial expectation. They were forced to prepare other options to deal with the increased number of resettlers: resettlement in upstream areas and promotion of fish breeding for those from Wonorejo and Saguling, respectively. In both cases, the plans were not very successful. In Wonorejo, the quality of farmland was poorer than the land that became submerged. In Saguling, the cash compensation was not enough for the resettlers to fully invest in fish breeding, and the planners failed to give resettlers the exclusive right to breed fish in the reservoir, attracting outsiders to the business. Intensive consultation and careful analysis of the resettlement plan after the surveys of the resettlers' choices should have been conducted in order to avoid or at least mitigate the problems.

Another important factor to be considered by project planners is the long-term approaches. While we know that satisfactory resettlement was attributed to successful child rearing and securing independent livelihoods for the second generation, it may also be influenced by events and/or the environment outside the control of resettlement planners. In the long run, we observe that it is virtually impossible to foresee all possible outcomes. Nonetheless, planners have to consider future uncertainty in the planning of resettlement. To do so, one may duly consider factors in a step-by-step manner. For example, one may first consider the future of the second generation, i.e., the resettlers' children. What kind of educational opportunities will be available to them? What kind of job opportunities will they have?

In sum, planners could explore the kind of life courses they may have within the project area or in the country. Considering these aspects does not mean that the preparation of these choices should be completed in the initial stages of resettlement. Suppose a high school was built in the newly developed area as part of the resettlement program; provision of the high school might easily seem to create an opportunity for higher education. However, this arrangement might not open new opportunities for children in the future if one considers the quality of an education based on inferior facilities compared to schools in the city and future job opportunities. Instead, if other forms of assistance were provided after resettlement, such as scholarships or dormitory facilities for resettlers' children to stay in the city for schooling, they may be quite helpful in satisfying the resettlers' needs. Another example might be fish breeding, as shown in Saguling. Negative incidences such as eutrophication would likely have occurred even if the reservoir was exclusively utilized by the resettlers. It is likely that the dam developers would not take any necessary countermeasures to make the business sustainable and profitable after some decades since completion of dam construction. For this kind of arrangement, it is not necessary to have perfect foresight or make complex predictions during the resettlement planning; instead, planners must have the will to take responsibility for resettlers' livelihoods for a certain period. We would like to call this attitude a necessary "far-sightedness." This far-sightedness necessary for a successful resettlement program may not consist of complete assessment of all future uncertainties and measures to address them, but just preparedness to take responsibility for the livelihoods of the resettlers.

The rent scheme adopted in the case of the Jintsugawa Dams is unique and presents many advantages, although it was adopted only as an urgent measure to begin dam construction as soon as possible. This approach is not adopted in Japan any longer, because national guidelines adopted later stipulate only cash compensation (Box 2.1). However, the rent scheme provides opportunities to deal with two important factors mentioned above: resettlers' choices and long-term perspectives. As long as the resettlers obtain rent, even if it is not enough to rehabilitate their lives soon after resettlement, then they will still have opportunities to improve their lives over time. By themselves, they will probably deal better with any problems that occur long after the resettlement. It is worth scrutinizing the potential effectiveness and applicability of rent schemes in other situations.

References

Bartolome, L. J., de Wit, C., Mander, H., and Nagraj, V. K. (2000). *Displacement, Resettlement, Rehabilitation, Reparation, and Development.* Cape Town: World Commission on Dams.

Cernea, M. (2003a). For a new economics of resettlement: a sociological critique of the compensation principle. *International Social Science Journal,* 55 (175), 37–45.

Cernea, M. (2003b). Personal communications, 12 December.

Chubu Electric Power Company Construction Department (1961). *Ikawa dam kouji-si* (Chronicle of Ikawa Dam Construction). Shizuoka: Chubu Electric Power Company.

Costa-Pierce, B. A. (1997). *From Farmers to Fishers: Developing Reservoir Aquaculture for People Displaced by Dams.* World Bank Technical Paper No. 369. Washington, DC: World Bank.

Costa-Pierce, B. A. (1998). Constraints to the sustainability of cage aquaculture for resettlement from hydropower dams in Asia: an Indonesian case study. *Journal of Environment and Development,* 7 (4), 333–368.

Djuangsih, N., Haryanto, E. T., Hastiawan, I., and Iskander, A. (1997). Eutrophication aspects in a man-made lake: a case study of Saguling Reservoir, West Java. Paper presented at workshop on water quality research and development, Bandung, February, IOE. Bandung, Indonesia: Padjadjaran University.

Egre, D. and Senecal, P. (2003). Social impact assessments of large dams throughout the world: lessons learned over two decades. *Impact Assessment and Project Appraisal,* 21 (3), 215–224.

Government of Japan (1953). *Dengen kaihatsu sonota ni yoru sonshitsu hosho yoko* (Principle for Compensation against Inundation for Power Generation). Government of Japan, April 14.

Gunawan, B. (1992). Floating net cage culture: a study of people involvement in the fishing system of Saguling Dam, West Java. MA thesis. Manila: Ateneo de Manila University.

Hanayama, K. (1969). *Hosho no riron to genjitsu* (Theory and Practice in Compensation). Tokyo: Keiso Shobo.

Hokuriku Denryoku (1962). *Hokuriku Denryoku Juu-nen-shi* (The First Ten Years of Hokuriku Denryoku). Chapter on Jintsugawa Dams No. 1, No. 2, and No. 3, pp. 168–175.

Hosoiri-mura (1987). *Hosoiri-mura Sonshi* (History of Hosoiri Village). Hosoiri-mura, Japan: Author.

Ikawa Village (1958). *Ikawa dam no kiroku* (Record of Ikawa Dam). Shizuoka: Shizuoka News Shinbunsha.

IOE (1979). *Environmental Impact Analysis of the Saguling Dam: Studies for Implementation of Mitigation of Impact and Monitoring.* Report to Perusahaan Umum Listrik Negara, Jakarta, Indonesia. Bandung, Indonesia: Padjadjaran University.

IOE (1985). Saguling Hydroelectric Power Plant. Paper presented at Workshop on the Evaluation of Environmental Impact Assessment Applications in ASEAN countries, 4–7 March. Bandung, Indonesia.

IOE-UNPAD and ICLARM (1989). *Development of Aquaculture and Fisheries Activities for Resettlement of Families from the Saguling and Cirata Reservoirs. Volume 2: Main Report.* Bandung, Indonesia and Manila, Philippines: IOE-UNPAD and ICLARM.

IPB (Institut Pertanian Bogor) (1985). *Draft Final Report—Analisa Dampak Lingkungan pada Proyek Tulungagung: Pembangunan Waduk Wonorejo* (Draft Final Report on Environmental Impact Assessment of Tulungagung Project: The Development of Wonorejo Dam). Bogor: IPB.

Japan Dam Foundation (2008). *Dam Binran 2008* (Dam Handbook, updated annually). Retrieved from http://damnet.or.jp/Dambinran/binran/TopIndex.html.

Manatunge, J., Contreras-Moreno, N., and Nakayama, M. (2001). Securing ownership in aquaculture development by alternative technology: a case study of the Saguling Reservoir. *International Journal of Water Resources Development*, 17 (4), 611–631.

Maruyama, Tamio. (1986). *Dam Hosyou to Suigen Chiiki Keikaku* (Dam Compensation and Head Water Regional Development Plan). Tokyo: Nihon Dam Kyoukai Kenkyu Bu.

Miyashita T. (1961). Sannoukai Dam to Jintsugawa Dai-ni Dam no Suibotsu Hosho (Compensation for Sannoukai Dam and Jintsugawa Dai-ni Dam). *Suiri Kagaku*, 5 (6), 165–179.

Miyata, S. and Manatunge, J. (2004). Lessons for sound policies in water resource management: evidence from households' decisions towards aquaculture in Indonesia. *International Journal of Water Resources Development*, 20 (4), 523–536.

Mizushigen Kaihatsu Koudan (1974). *Kusaki dam Kensetsu ni okeru Suibotsu Iten Setai no Seikatsu Saiken Jittaichousa Houkokusyo.* (Study Report on the Current Status of the Rehabilitation of Livelihood by Submerged Households in the Case of Kusaki Dam Construction). Tokyo: Mizushigen Kaihatsu Koudan.

Nakayama, M. (1998). Post-project review of environmental impact assessment for Saguling Dam for involuntary settlement. *International Journal of Water Resources Development*, 14 (2), 217–229.

Nakayama, M. and Furuyashiki, K. (2008). From Expropriation to Land Renting: Japan's Innovations in Compensating Resettlers. In Michael M. Cernea and Hari Mohan Mathur (eds.) *Can Compensation Prevent Impoverishment?* Delhi: Oxford University Press, pp. 357–374.

Nakayama, M., Gunawan, B., Yoshida, Y., and Asaeda, T. (1999a). Resettlement issues of Cirata Dam project: a post-project review. *International Journal of Water Resources Development*, 15 (4), 443–458.

Nakayama, M., Yoshida, T., and Gunawan, B. (1999b). Compensation schemes for resettlers in Indonesian dam construction projects: application of "Japanese soft technology" for Asian countries. *Water International*, 24 (4), 348–355.

Nihon Dam Kyoukai (The Foundation for Japan Dam Association) (1978). Sameura Dam Kensetsu niyoru Chiiki Syakai no Henka to Suibotsu Itensya he no Eikyo ni tsuite (Influence to the Community Transformation and Resettlers by Sameura Dam Construction). In *Dam kensetsu to suibotsu hosyou* (Dam Constructions and Compensations Measures). Tokyo: Nihon Dam Kyoukai.

Nippon Koei (2002). *Completion Report Volume II: Supporting Report Package-1B Infrastructures at Resettlement Site*. Jakarta: Directorate General of Water Resources Development, Ministry of Public Works.

Okada, T. (2004). *Wonorejo Multipurpose Dam Construction Project* (Adobe Digital Editions Version). Retrieved from www.jica.go.jp/english/our_work/evaluation/oda_loan/post/2005/pdf/2-05_full.pdf.

PLN (1989). *Rencana pengelolaan dan pemantauan lingkungan (RKL & RPL) PLTA Saguling* (Planning for Environmental Management and Monitoring of Saguling Hydroelectric Power). Jakarta: PLN.

Schaap, B. (1974). *Report on a Household Study of Nam Ngum Reservoir Evacuees with Recommendations for a Programme of Action*. Bangkok: Mekong Committee.

Seta gun Azuma Mura (Azuma village, Seta County). (1998). *Seta gun Azuma Mura Shi- Tsushi- hen* (General History of Azuma Village, Seta County). Seta gun Azuma Mura Shi Hensan Shitsu (Editing office in Azuma Village, Seta County). Gunma, Japan: Seta gun Azuma Mura.

Sinaro, R. (2007). *Menyimak Bendungan di Indonesia (1910–2006)* (Review on Indonesian Dams (1910–2006)). Jakarta: Bentara Adhi Cipta.

Soemarwoto, O. (1990). Introduction. In B. A. Costa-Pierce and O. Soematwoto (eds.) *Reservoir Fisheries and Aquaculture Development for Resettlement in Indonesia*. ICLARM Tech. Rep., Philippines: ICLARM, pp. 1–6.

Souri-fu Shigen Chousa-kai (Resource Research Committee for the Prime Minister's Office) (1954). *Mizushigen No Kaihatsu Ni Tomonau Hoshou Jirei-Shuu* (Some Cases of Compensation Associated with Water Resources Development). Chapter on Jintsugawa Dams Nos. 1 and 2. Tokyo: Souri-fu Shigen Chousa-kai.

Sutandar, Z., Costa-Pierce, B. A., Iskandar, R., and Hadikusumah, H. (1990). The aquaculture resettlement option in the Saguling Reservoir, Indonesia: its contribution to an environmentally-oriented hydropower project. In R. Hirano and I. Hanyu (eds.) *The Second Asian Fisheries Forum*. Manila, Philippines: Asian Fisheries Society, pp. 253–258.

Suwartapradja, O. S., Arifin, T., Kanum, A., Ansor, and Djumari. (1985). *Pemantauan sosial-ekonomi budaya penduduk pindahan dari bawah ke atas genangan PLTA Saguling* (Monitoring on Socioeconomic and Cultural of the Displaced People). Pusat Penelitian Sumber Daya Alam dan Lingkungan. Bandung, Indonesia: Universitas Padjadjaran.

Takashima, G. (1956). Atarashii Mura-dukuri—Shizuoka-ken Abe-gun Ikawa-mura (New Village Building—Shizuoka Prefecture Abe district Ikawa village). *Fumin* (Prosperous Farmers), 28 (1), 67–71.

Takesada, N. (2006). Lessons from the Japanese experience in involuntary resettlement for dam construction: Case of New Village Building. In Wing-Huen Ip and Namsik Park (eds.) *Advances in Geosciences Volume 4: Hydrological Science (HS)*. Singapore: World Scientific Publishing Co. Pte. Ltd, pp. 143–148.

Box 2.1 Compensation schemes for dam submergence zones in post-war Japan

Japan's Compulsory Purchase of Land Act (1951) established the basis for compensation systems for private land and property inundated by dam construction. Based on this act, the "Guidelines on Standards for Compensation for Losses Associated with the Acquisition of Land for Public Purposes" (approved by a 1962 Cabinet decision, hereafter "Guidelines") set standards for compensation in the case of land expropriation. The Guidelines established the principle that losses of land and property due to inundation would be compensated financially. Compensation in the form of actual land or property is acceptable "where circumstances permit," when "genuinely inevitable," but this approach really only provides a code of conduct for "best effort." Until the Guidelines were established, compensation in the form of actual land was already being given to people who had lived in submergence zones (Takesada 2006), but after the Guidelines came into effect, compensation was in principle only done by financial payments.

On the same day the Guidelines were approved, a Cabinet decision was reached on a "Memorandum on the Implementation of Guidelines on Standards for Compensation for Losses Associated with the Acquisition of Land for Public Purposes." It established implementation policies for the Guidelines. Stating that "no longer will compensatory measures be taken as was done in some previous cases for hard-to-define items such as compensation for psychological loss and financial incentives to encourage cooperation, etc." This document indicates that if compensation is given based on the Guidelines, it is not necessary to provide compensation for the right to livelihood, and that only property rights are subject to compensation. However, the Memorandum also states that "where required, an effort should be made to take measures to help secure land and buildings in order to restore livelihoods, or to provide job placement or guidance," leaving the door open for developers to assist in the restoration of livelihoods of dam-affected households (Section 2 of the Memorandum). Based on this thinking, after the approval of the Guidelines, governments (mainly local governments) promoted assistance for the restoration of livelihoods of resettled people. With some additional laws for reconstruction and redevelopment of affected community facilities introduced in 1970s, the Guidelines are still in effect and have provided the main principles for compensation in Japan. Considering the economic situation in Japan in the 1950s and 1960s, the main purpose of the Guidelines was to allow public works to be implemented as smoothly as possible, without causing trouble for the

(continued)

(continued)

bodies executing the projects, and *not really* to assist the smooth restoration of the affected people's livelihoods (Hanayama 1969).

In the case of submergence of public facilities, the "Guidelines on Standards for Public Compensation Associated with the Implementation of Public Works Projects" were approved by Cabinet decision in 1967, recognizing that there could be public compensation for public facilities submerged by developers. While these guidelines are based on the principle of financial compensation, what differs from compensation for private property is that in-kind compensation is also acceptable "in cases where it is considered to be practical from the technical or economic perspective." Based on these guidelines, it was officially permissible for dam developers to provide compensation for losses by building things such as schools and roads for residents in resettlement areas.

As the country became more democratic and citizens became politically aware, the promotion of dam construction projects became more difficult. A 13-year campaign from 1958 through to 1971 against construction of the Matsubara-Shimouke Dam in rural Kyushu had a huge and lasting impact on the nature of dams and other public works projects in Japan. In fiscal year 1969, the Ministry of Construction established the "Rehabilitation Measures Fund" within its Dam Construction Fund, and, in 1973, the Act on Special Measures concerning Measures Related to Water Resources Areas (hereafter "ASM") was passed.

The purpose of the ASM was "to improve the stability and welfare of livelihoods of affected residents," and to "promote construction" (Article 1). The ASM was designated for dams specifically, and stipulated that the relevant government authorities should make various infrastructural improvements in the areas surrounding dams. In addition, the law contains express provisions stating that the relevant authorities are to provide assistance for resettlers to obtain land and buildings, as well as job placement, guidance, and training. Dams can be designated under the ASM if 20 or more households are to be submerged, or at least 20 ha of farmland (in Hokkaido, at least 60 ha). If a project is so designated, the prefectural governor must create a watershed improvement plan to be approved by the prime minister. After this Act entered into force, 112 dams were designated under the ASM by the end of fiscal year 2011 (MLIT 2013).

In 1976, the "Watershed Countermeasures Fund" was established as a supporting program for the ASM. Local governments in areas upstream as well as downstream that benefit from dams are to contribute to the Fund, which serves as a financial mechanism to provide interest assistance for dam-affected households in securing resettlement sites, to provide livelihood consultation personnel, to build or improve roads for livelihood purposes, and to facilitate exchanges between communities upstream and downstream.

In 2001, the Compulsory Purchase of Land Act was extensively revised for the first time in 35 years, this time incorporating livelihood-rehabilitation measures similar to those stipulated in the ASM. The revised act differs in a significant way from the ASM, which calls for a best-effort basis by developers for livelihood rehabilitation, by the fact that it requires livelihood restoration to be offered to people displaced by any public works project.

(*Naruhiko Takesada*)

References

Hanayama, K. (1969). *Hosho no riron to genjitsu* (Theory and Practice in Compensation). Tokyo: Keiso Shobo.

MLIT (Ministry of Land, Infrastructure and Transport) (2013). *Suigen chiiki bijon* (Water Resource Area Vision). Retrieved from www.mlit.go.jp/river/kankyo/main/kankyou/suigen/index.html.

Takesada, N. (2006). Lessons from the Japanese experience in involuntary resettlement for dam construction—Case of New Village Building. In Wing-Huen Ip and Namsik Park (eds.) *Advances in Geosciences Volume 4: Hydrological Science (HS)*. Singapore: World Scientific Publishing Co. Pte. Ltd, pp. 143–148.

Box 2.2 Indonesia's transmigration

Transmigration is Indonesia's resettlement program (also known as TP), intended to solve problems of density and unequal distribution of population across the country's islands. More than a century has passed since its first implementation in 1905. Today, transmigration has transformed itself to function as a national strategy of regional development.

As the largest archipelagic nation in the world, Indonesia has more than 17,000 islands and a total area 1.9 million km². According to the latest census, the population has reached 237.6 million, bringing population density to 124 persons/km² (BPS 2013).

The average population density is relatively low, but unequal across the islands. Java is the most populated island with only 6.8 percent of the total area of Indonesia, but 57.5 percent of Indonesia's total population. The remaining islands account for 93.2 percent of the nation's total area, but only 42.5 percent of its population. The unequal population distribution results in a large disparity in population density, with

(*continued*)

(continued)

more than 1,000 persons/km² on Java, but fewer than 50 persons/km² elsewhere.

The disparity of population density across the islands is actually not a new demographic phenomenon. The Dutch colonial government long recognized the problems of low productivity due to the disparity of population density. Agricultural expansion in Java was limited by the availability of land, while outside Java it was limited by the supply of labor. In order to reduce population density in Java and increase the supply of labor outside Java, the government introduced a program known as *kolonisasi*, or colonization. *Kolonisasi* was designed to reduce population density in Java by relocating families to other islands. A person participating in the program was called *kolonis*. In November 1905, the *Kolonisasi* program resettled 155 families from Java to Lampung on the island of Sumatra, the first relocation in the Indonesia's history of transmigration (Hardjono 1982; Syamsu 1960). The program continued to relocate families away from Java to other islands until 1942, resettling 173,959 persons from 1905 to 1941 (Kemenakertrans 2012).

After the proclamation of independence in 1945, the government of Indonesia transformed the program from *Kolonisasi* to *Transmigrasi* or transmigration. The term *kolonis* now became *transmigran* or transmigrant. Between 1905 and 2013 the government of Indonesia successfully resettled 2,138,312 families (7,936,651 persons) from Java Island to 5,885 resettlement villages on other islands (Kemenakertrans 2012).

The TP has continued to develop. It is now no longer simply a program to move people from areas of higher to lower population density. The government enhanced the program to support food security and housing supply, alternative energy development, regional investment, and national defense in outer and bordering territories, and to contribute a solution to the problems of unemployment and poverty (Kemenakertrans 2012).

As a development strategy, the TP involves several goals. First, it is designed to equalize the distribution of population by resettling families from more- to less-densely populated regions, thereby improving labor productivity in the region of origin and land productivity at the destination. Next, the program is intended to strengthen local and national defense and security, with the belief that reducing population density in some regions has the potential to reduce the risk of crime. Finally, the program is intended to raise the standard of living by developing individual and collective capacity, by providing houses and agricultural land.

In implementation, there are three approaches in practice. The first is general transmigration, implemented under the coordination of

Ministry of Labor and Transmigration, with full government sponsorship and financing. The second is voluntary and autonomous, based on the personal initiative of transmigrants, for whom the Ministry provides guidance, supervision, and supporting facilities. The third is *bedol desa*, which means an entire village is relocated, with complete systems from the original village, including social, cultural, economic, and administrative systems. *Bedol desa* transmigration is applied to accommodate displaced families in villages damaged by natural disasters or development projects. Relocation for the Koto Panjang Dam development followed the *bedol desa* approach.

To strengthen the institutional basis of the TP, the government issued Law Number 29 in 2009. The law positions the TP as an integrated regional development strategy to attract wider participation from local governments, local communities, and business firms. The TP is expected to function as an engine of regional economic development, and to promote the development of small, autonomous cities and local economic competitiveness. At the same time, it is expected to reduce economic disparities between regions, and to strengthen the linkages between urban and rural economies. With more local autonomy, the dominant role of the central government is to be reduced. Local governments are encouraged to develop their own initiatives in allocating local areas for transmigration sites and promoting the transmigration areas as integrated regional development centers.

Meanwhile, the TP has sometimes failed to improve the situation of transmigrants. Many transmigrants actually received less agricultural land than they were promised. The soil and climate of transmigration areas were sometimes not as productive or beneficial as in Java and Bali compromising their chances of life rehabilitation. The program has been also accused of accelerating the deforestation of tropical forests.

(Syafruddin Karimi)

References

BPS (2013). *Statistik Indonesia 2012*. Jakarta: Badan Pusat Statistik.

Hardjono, J. (1982). *Transmigrasi: dari Kolonisasi sampai Swakarsa*. Jakarta: Gramedia.

Kemenakertrans (2012). *Rencana Pembangunan Jangka Panjang Bidang Ketenagakerjaan dan Ketransmigrasian Tahun 2010–2025*. Jakarta: Kementerian Tenaga Kerja dan Transmigrasi.

Syamsu, A. (1960). *Dari Kolonisasi ke Transmigrasi*. Jakarta: Djembatan.

3 Proper implementation of original resettlement programs

Introduction

In this chapter, we present numerous issues observed in cases in Indonesia, Lao PDR, Sri Lanka, and Japan. Resettlement programs were not always implemented properly, with some being delayed or not implemented according to original plans. In cases of land-for-land compensation, developers sometimes obtained insufficient land to meet the needs of resettlers. Land in resettlement areas provided for resettlers was not always prepared as promised or was much less productive than their now-submerged farmland. In cases of cash compensation, resettlers often were only able to purchase smaller land plots due to increases of land price during negotiations for compensation. As a result, in land-for-land and cash-for-land schemes, the income of resettlers decreased and/or became unstable. Lengthy negotiations did not necessarily result in a good resettlement outcome. Changes in the social and economic situation can also negatively affect the outcomes.

In the cases of the Bili-Bili Dam and the Koto Panjang Dam in Indonesia, the resettlement areas were not developed as had been promised, and the resettlers were faced with serious hardship for years after resettlement. The resettlers from the Nam Theun 2 Dam in Lao PDR were provided with only a small piece of land, insufficient to sustain their livelihood. Resettlers from the submerged area of the Kotmale Dam in Sri Lanka were faced with instability of income due to a sharp decline in the international price of rice, while the Sameura Dam in Japan demonstrates the undesirable consequences of lengthy negotiations. In that case, resettlers received payment only for their material losses, and were not able to obtain any additional compensation to secure their livelihoods after relocation.

Bili-Bili Dam (Indonesia)

The Bili-Bili Multipurpose Dam was constructed on the Jeneberang River in the district of Gowa, 31 km from the city of Makassar on the Indonesian island province of South Sulawesi. Construction was completed in 1997, making it the first large dam in eastern Indonesia. It was constructed primarily for flood control, municipal and irrigation water supply, and hydropower. The dam construction submerged an area of 2,802 ha, and 2,085 households had been resettled by the year 2000. Relatively rich households (1,494 in total) were able to afford new land and/or build a new house with their compensation money from the dam project and move to a new location within the district of Gowa (1,079 households) but close to the reservoir, or to urban areas (415 households) such as the city of Makassar. The rest (591 households) were relatively poor, and many were landless; they were not able to obtain enough compensation to buy a house in Gowa, and chose to take advantage of an offer of free farmland and a house in the districts of Mamuju and Luwu under Indonesia's TP (Transmigration, see Box 2.2). Both Mamuju and Luwu are located more than 400 km away from the original settlement in question. Within a few years of relocation, however, it was discovered that many of the households resettled to Mamuju and Luwu had dropped out of the TP, and were thought to have returned to the areas in which they used to live, close to the reservoir. (The reasons for returning are analyzed in Chapter 5.)

Resettlement to transmigration areas

The resettlement was carried out in the following manner: the Directorate General of Water Resources of the Ministry of Public Works, in cooperation with the local government—in this case the district of Gowa in the province of South Sulawesi—conducted all activities, including land acquisition and evacuation of residents. After identifying 2,085 households that would need to be resettled, as well as the relevant titles to land, rice fields, and plantation properties, they began conducting a series of explanatory meetings with residents in 1987 in order to gain agreements on land and property acquisition.

In 1988, the people and the government reached an understanding that included the agreement that all types of compensation for land, houses, plants, and other properties were to be paid in cash, and that all resettlers were free to choose the location of their new settlement. It was confirmed that all of them did receive their cash compensation (PPLH Unhas 1998). The Ministry of Transmigration agreed to allocate land for 200 households in Luwu and 391 households in Mamuju. The villages in Luwu and Mamuju were developed as TP projects exclusively for the resettlers from Java and Bali. It was an exceptional measure to accommodate Bili-Bili resettlers. The company in charge of the dam construction project also agreed to support the construction of infrastructure in those two transmigration areas.

As mentioned, among the 2,085 households that relocated, 1,079 moved within Gowa to the vicinity of the reservoir, and 415 moved to urban locations such as Makassar. Most were relatively rich and had owned land in the newly submerged area, which meant that the compensation money they received could go towards buying or building a new house. Wealthier households chose to move to Makassar to seek better jobs. (Their life in the urban area is analyzed in Chapter 6.) Those who owned cultivated land that would not be submerged remained in Gowa.

As mentioned, the rest of the households (591) chose to move to the transmigration areas, with 200 going to the district of Luwu and 391 to the district of Mamuju. The reasons for their choice were as follows: (1) they were poor (some were landless) and unable to afford a house in the vicinity of the reservoir; (2) they wanted to save their compensation money, as the transmigration land and house were provided free; (3) they obtained a living support subsidy for one year; and/or (4) they wanted to explore the opportunity for improved living conditions.

The destinations of households under the TP depended on the year in which they moved, due to the availability of new living quarters. The first 200 households dispatched from Bili-Bili were to go to Luwu, but only 141 of the registered households actually arrived in Malangke IV, one of the transmigration units in Luwu, in 1990. The remaining 59 households, who had registered as transmigrants, decided at the last moment to stay in their neighborhood. The last households were dispatched in 1995 to the Mora II Transmigration Unit in Mamuju. Most of the transmigrants from Bili-Bili moved to Malangke IV in Luwu and Tommo V in Mamuju, because other units were already inhabited by transmigrants from Java and Bali.

Conflicts with local inhabitants in Luwu

The resettlers at the Malangke IV transmigration site in Luwu faced a number of problems. For one, the designated resettlement area in Luwu was inhabited not only by newcomers from Bili-Bili, Java, and Bali but also by local people already settled there for a long time. The real estate designated for new homes for the transmigrants from Bili-Bili rose in value, because the Ministries of Transmigration and Public Works built additional infrastructure exclusively for the resettlers from Bili-Bili. This provoked the envy of the original inhabitants. They insisted that they should get the same rights as those obtained by the transmigrants and they protested land-allocation plans for the transmigrants, including the land allocated for those from Bili-Bili.

The local government also intended to allocate a natural sago palm forest to the transmigrants, but it was communal land and had been traditionally utilized by the local people. Further, cutting trees or cultivating land there was forbidden under a traditional local custom. As a result of the conflict, most of the resettlers were not provided the farmland they had been promised, and ended up working at the nearby orange plantation. In the beginning, the income for plantation workers was satisfactory, but the orange plantation was damaged by a viral infection in 1997. Many workers thus lost their jobs by 1998. The head of the village was, understandably, always more sympathetic to the local inhabitants, and even attempted to seize the 59 unclaimed transmigrant houses for them.

At the same time, a good access road to the Bili-Bili area facilitated the return of transmigrants from their new resettlement houses to areas near their original homes. According to a national regulation, transmigrants who abandon their new land for more than three months lose their title to that land. The head of the village actively voided the rights of the transmigrants who left their new homes. This created unfair conditions for the transmigrants, but they did not know where to file complaints since this deal was not directly attached to the dam construction project. As a result, resettlers who had been landowners in Bili-Bili, and had enough money, left Malangke. By 2004, the number of resettled households there had decreased from 141 to 41. Most of the remaining resettlers were originally landless in Bili-Bili and therefore had obtained little compensation. In 2004, the Ministry of Labor and Transmigration and the Luwu District government admitted to the failure of the arrangement in Malangke IV, and provided the households remaining there with a new house and farmland elsewhere in the same district.

According to interviews with eight resettled households in Sepakat of Luwu in 2011, all of them first resettled in Malangke IV, where resettlers were given only 0.25 hectares of land each despite the government's promise of two hectares each, and they were once again forced to relocate to Sepakat, where they were given one hectare of land. This is still only half of what was promised by the TP, but they managed to make it work. All resettlers suffered from water shortages caused by lack of infrastructure.

Satisfaction of resettlers in Mamuju

The Tommo V transmigration site, renamed Campaloka Village after management was handed over from the Ministry of Transmigration to the local Mamuju District government, was established in the 1991/1992 fiscal year. The total population was 2,145 people in 636 households. It is located on a floodplain, and work on a cacao plantation is the main source of income for residents. This area, where rice cultivation was formerly the main source of income, was developed as a condition of the first year of resettlement, and cultivating cacao provides more stable incomes to farmers than rice (PT Andel Persada 2006). Unlike in Luwu, the transmigration area in Mamuju is relatively distant from where the local inhabitants live. They were able to elect their own village head, who began to work seriously on behalf of the villagers.

Some of the infrastructure in Mamuju, including roads and drainage, was not fully developed in the early stages of receiving transmigrants (around 1991). Prior to the full development of this infrastructure, the resettlers and other transmigrants experienced significant hardship. They were often forced to travel long distances for crop trading, and the lack of flood-control infrastructure often resulted in harvest failures. Those who were unable to bear the hardships left Mamuju. Tommo V hosted 194 resettled households (including transmigrants from other islands); however, 115 resettled households (59.3 percent) left, leaving only 79 resettled households (40.7 percent) residing there until 2007. Resettlers who were able to overcome the early difficulties were generally satisfied with their lives and stated that they wished to continue living in the area (Yoshida et al. 2010).

In-depth interviews were conducted with 14 of the 79 transmigrant households from Bili-Bili (Table 3.1). In general, most perceived their post-resettlement living conditions as "good." All respondents reported marked improvements in living conditions, and they found them better in comparison to other resettlement areas also inhabited by resettlers from Bili-Bili. Twelve households were satisfied with their livelihoods as farmers, and 11 households were satisfied with their economic situation.

Satisfaction of resettlers adjacent to Bili-Bili reservoir

Most of the resettlers who chose to remain in the area adjacent to the reservoir moved within the district of Gowa. Unlike the transmigration areas, however, only lower-grade infrastructure and facilities were available in Manuju in Gowa. In-depth interviews were conducted in four villages in Gowa in order to identify the living conditions of the resettlers before and after resettlement, and to identify whether or not they belonged to the poor households' group, in comparison with other households in the same village, so as to avoid flawed data.

In order to establish the criterion for identifying a "poor" household, the prosperity ranking method was used with 100 inhabitants, including both non-resettlers and resettlers, randomly sampled from four villages in Gowa. Among the 100 households, 91 had an income below IDR 300,000 per month (about USD 33 per month in 2007), and only 10 of the 91 households had farmland. Based on this information, households with an income below IDR 300,000 per month were defined as being poor.

Table 3.1 Perceptions of resettlers in Tommo V (Mamuju District) on their living conditions and infrastructure (14 interviewed in 2007)

Category	Satisfied	Unsatisfied	No comment
Occupation	12	0	2
Location	14	0	0
Security	13	1	0
Economic condition	11	0	3
Housing	9	2	3
Domestic water availability	0	14	0
Road access	13	1	0
Mosque	12	2	0
Health care services	14	0	0

Source: Authors.

Next, 22 resettlers were identified from among the households in the sample group of 100 and interviewed. These were not households returning from the transmigration areas; they had moved directly from the area now submerged. The age of respondents ranged from 30 to 73 years. Two were women, 20 had moved between 1990 and 1993, while one household moved in 1987 and another in 1996. In general, the resettled households did not change their main occupation, i.e., all of them were still farmers who owned their own rice paddy field, and their incomes had not changed much. Fourteen households had incomes between IDR 300,000 and IDR 1,000,000 per month after the resettlement.

The results of a study conducted by the authors in 2004 are summarized in Table 3.2. It is obvious that the resettled households became wealthier after moving and also richer than those who were original Manuju residents. For instance, there was no electricity before the dam's construction. The electricity supply system was set up afterwards as part of the compensation package provided to the surrounding community. In terms of monthly electricity bills, 19 of the 22 resettled households paid more than IDR 30,000, while 80 percent of all inhabitants paid less than IDR 30,000. This is reflected in the better condition of and higher number of household appliances owned by the resettled households in comparison to the original households. Most of the resettled households owned a television and other electrical appliances, while 70 percent of original inhabitants has none.

Table 3.2 Result of a study in 2004 on the conditions of 22 resettlers before and after resettlement to Gowa compared to 100 randomly sampled inhabitants in Gowa

Indicator	100 original inhabitants in Gowa District	22 resettlers	
		Before resettlement	*After resettlement*
Income			
IDR 300,000–IDR 1,000,000	9	13	14
< IDR 300,000	91	9	8
Main occupation			
Farmer (land owner)	10	17	17
Tenant	50	5	5
Other (sand mining, etc.)	40	0	0
Type of house			
Wooden	85	19	0
Half-brick/wooden	15	3	6
Brick	0	0	16

(continued)

Table 3.2 (continued)

Indicator	100 original inhabitants in Gowa District	22 resettlers	
		Before resettlement	After resettlement
Electricity bill payments			
No electricity	0	22	0
< Rp 30,000/month	80	0	3
> Rp 30,000/month	20	0	19
Appliances			
No electric appliances	70	21	0
Radio only	20	1	11
Radio and TV or refrigerator, etc.	10	0	11
Sources of education financing			
Self-financed	65	21	13
Scholarship	12	1	9
Other	13	0	0

Source: Authors.

The prosperity-ranking indicators show that many resettled households did not belong to the poor household category. Their living standards were much higher, so they belonged to either the moderate or rich household category. Of the 22 resettlers, 20 said they used their compensation money from the dam project's land acquisition to buy new land and/or to build new houses. Some of them invested the money to start small businesses.

The resettlers enjoyed a better living standard in their current villages. Most of the area adjacent to the reservoir and both transmigration areas have better transportation access and newly provided electricity after the relocation, but most of the areas still lack clean water during the dry season. While all indicators show that the resettlers' living conditions were much better than the poor households, they tended to under-report their living standards in general. Both poor households and resettlers complained that the dam/reservoir did not bring more obvious direct benefits; most of them held the opinion that the reservoir itself has had neither positive nor negative direct impacts on their living conditions.

Koto Panjang Dam (Indonesia)

The Koto Panjang Dam was planned and constructed in the middle of Sumatra Island, on the border between the provinces of Riau and West Sumatra, by PLN, the state-owned power company, with concessional yen loans provided by the OECF, Japan. Its objective was to provide stable electrical power, both to improve the regional electrification rate and to meet increasing electricity demand in the region. An area of 124 km² was submerged due to the reservoir. The residents of eight villages in Riau and two villages in West Sumatra had to be relocated, and 4,886 households were displaced, totaling 16,954 people (JBIC 2004).

Under the resettlement program, every household was to receive the following benefits in the resettlement villages: (1) two hectares of productive rubber plantation; (2) 0.4 ha of land to grow food; (3) 0.1 ha of land for a 36 m² house with a courtyard; (4) living allowance for two years; and (5) monetary compensation for all properties given up as a result of dam construction. With these provisions for compensation, villagers were to choose new resettlement villages prepared by the government of Indonesia, mainly in the vicinity of the reservoir (i.e., near the original villages). Resettlers also had an option to live and work on an existing palm oil plantation. Further, the resettlement program had provisions for the following public facilities in each resettlement village: (1) drinking water supply; (2) a school; (3) a health care center; (4) a mosque; and (5) a marketplace.

This resettlement program aimed to sustain, at the very least, or improve the living standards of resettlers compared to conditions before resettlement. Since smooth and early recovery of livelihood was thought to be the most critical issue in resettlement during the planning stage, the program tried to provide resettlers with security in continuing their major occupation and income-earning activities. There was also consideration of a transition period until rubber plantations were ready for harvesting. For the transition period, the program planned for food provision and the introduction of cash crops other than rubber, such as chili, which has a relatively quick growth cycle, in coordination with the concerned ministries.

It is important to examine whether the implementation of the resettlement program mentioned above led to the recovery of resettlers' livelihoods to the levels "before resettlement." However, securing a solid answer to this question is neither easy nor straightforward. Different post-project surveys came to different conclusions.

For this study, the two resettlement villages of Tanjung Balik and Tanjung Pauh in West Sumatra were selected, related to the Koto Panjang Dam construction project. In total, 800 households (450 for Tanjung Balik and 350 for Tanjung Pauh) relocated in 1993 to the present villages, which were established by the resettlement program at a distance of about 15 km from the original villages. A post-project survey by the authors in 2004 (Karimi et al. 2005) revealed that Tanjung Pauh is one of the "least successful" among resettlement villages in both Riau and West Sumatra provinces. Even so, 60 percent of respondents in Tanjung Pauh experienced an increase in income after relocation, and among them 22.2 percent possessed motorcycles, where the figure for motorcycle ownership before relocation was 6.7 percent. These figures suggest that many households in Tanjung Pauh experienced livelihood improvements after relocation.

Conversely, a post-project survey carried out in 2002 by the Japan Bank for International Cooperation (JBIC, established through a merger of the OECF and the Export Import Bank Japan), a partial financer of the Koto Panjang Dam project, found that 71.6 percent of households in Tanjung Pauh felt that their living conditions were worse than before resettlement and that only 18.6 percent of households enjoyed better living conditions now (JBIC 2004). The situation was more or less similar in Tanjung Balik, where 66.5 percent of households found their living conditions worse than before, and only 10.3 percent of households experienced improved living conditions (JBIC 2004). This survey suggests that the resettlement program failed to work, at least for the two Koto Panjang resettlement villages in West Sumatra.

The differences between findings of the post-project surveys mentioned above could be attributed to the fact that one of them was carried out by JBIC. On the one hand, the respondents might have been motivated to present a "dark picture" with a view to seeking additional assistance from JBIC. On the other hand, perhaps respondents did not expect any potential monetary gain from the post-project survey carried out by the authors in 2004 (Karimi et al. 2005), leading to a "brighter picture" as the survey outcome.

In order to eliminate bias caused by resettlers' expectations, the authors conducted a comparative study between resettlement villages and the surrounding region, and examined if the living conditions of resettlers were either superior or inferior to those experienced by others (non-resettlers) in the region.

This study aimed at:

1   identifying the poor on the basis of property ownership by households;
2   analyzing the condition of poor households in West Sumatra's Koto Panjang resettlement villages;
3   relating the characteristics of poor households in resettlement villages to the implementation of the local development program; and
4   comparing the conditions of poor households in resettlement villages with those in West Sumatra as a whole.

For these objectives, this study used data from the Central Bureau of Statistics (CBS) of the government of Indonesia. The CBS had conducted a broad survey in 2005 to identify the characteristics of poor households, using minimum expenditure per capita to define the poverty line. This information was originally needed to distribute cash compensation to poor households that were negatively affected by a reduction in a government oil subsidy. The census covered all areas in Indonesia, including the Koto Panjang resettlement villages. The data available from the CBS are the most comprehensive for an analysis of the conditions of poor households, so our study utilized the data to analyze the conditions in West Sumatra's Koto Panjang resettlement villages.

For the purpose of identifying the poor, our study used the national poverty line of Indonesia: a monthly per capita expenditure less than IDR

175,000 identified an individual as poor (IDR 9,000 was equal to approximately one USD in 2005). We estimated the number of poor in the population based on the census data. Interviews were carried out with the village heads in 2005 to identify poor households in the locality. Each identified poor household was visited, and the interview verified whether the livelihood of the household matched the poverty criterion set by the CBS. Using the same threshold as the national government, we identified a household as poor if its total assets were less than IDR 500,000 (including cash, gold, cattle, motorcycle, color television, etc.).

Findings

Using this standard, we classified 211 households (131 households in Tanjung Balik and 80 households in Tanjung Pauh) in West Sumatra's Koto Panjang resettlement villages as poor. The proportion of poor households in Tanjung Balik was higher than in Tanjung Pauh. The poor account for 24 percent of resettled households (884 households in total in both villages during our study). This figure is significantly higher than the average poverty rate in West Sumatra of 5.5 percent (BPS 2006).

Table 3.3 shows the characteristics of poor households in the two Koto Panjang resettlement villages, in comparison with poor households in West Sumatra as a whole. About a third of poor households in the two villages

Table 3.3 Characteristics of poor households in the two Koto Panjang resettlement villages and in West Sumatra as a whole (percent)

		Village			
		Tanjung Balik	*Tanjung Pauh*	*Average (two villages)*	*West Sumatra*
Floor size per capita	Less than 8 m²	35.9	40.0	37.4	56.0
	More than 8 m²	64.1	60.0	62.6	44.0
Types of floor construction	Soil	16.0	23.8	19.0	67.7
	Cement	84.0	76.3	81.0	32.2
Type of house wall	Bamboo	71.8	83.8	76.3	75.8
	Concrete	28.2	16.3	23.7	24.2
Toilet	Public	71.0	75.0	72.5	90.5
	Private	29.0	25.0	27.5	9.5
Source of drinking water	Unprotected water well	29.8	17.5	25.1	87.1
	Piped water	70.2	82.5	74.9	12.9

(continued)

Table 3.3 (continued)

		Village			
		Tanjung Balik	Tanjung Pauh	Average (two villages)	West Sumatra
Lighting	Non-electric	48.9	37.5	44.5	63.7
	Electric	51.1	62.5	55.5	36.3
Cooking fuel	Wood	100.0	97.5	99.1	90.7
	Kerosene	–	2.5	0.9	9.2
	Gas/electricity	–	–	–	0.1
Consumption of beef, chicken, or milk (per week)	Never	95.4	37.5	73.5	88.4
	Once	4.6	62.5	26.5	11.3
	Twice or more	–	–	–	0.4
Number of meals for household members (per day)	One	0.8	–	0.5	6.0
	Two	50.4	42.5	47.4	62.8
	More than two	48.9	57.5	52.1	31.3
Purchase of new clothes (per year)	Once	61.1	82.5	69.2	70.4
	More than once	0.8	2.5	1.4	1.7
Access to health center	Yes	83.2	68.8	77.7	44.7
	No	16.8	31.3	22.3	55.3

Source: BPS (2006).

lived in houses with a floor size less than eight m² per capita, unlike the situation of poor households in West Sumatra as a whole. However, houses with bamboo walls, public toilets, wood cooking fuel, and the absence of weekly beef or chicken consumption dominated the conditions of poor households in the Koto Panjang resettlement villages, as they did among poor households in other areas of West Sumatra.

Poor households in Koto Panjang resettlement villages had good access to electricity and drinking water. More than 50 percent of poor households in the resettlement villages utilized the supply of electricity, while more than 60 percent of poor households in other areas of West Sumatra still used non-electrified lighting. Almost 75 percent of poor households in Koto Panjang resettlement villages had piped drinking water, while more than 85 percent of poor households in the other areas were still using unprotected water wells.

Table 3.4 shows the main occupations of poor households. Unemployment is evidently related to poverty. The proportion of households that were unemployed was higher in Tanjung Balik and Tanjung Pauh than in the province as a whole. Poor households mostly worked in plantation and rice agriculture in Tanjung Pauh and Tanjung Balik, respectively.

The plans were to allocate two hectares of rubber plantation for each resettled household under the Koto Panjang resettlement program. Productive rubber plantations were intended to sustain the livelihoods of 800 resettling households, but the central government failed in implementation of the plans. The failure of rubber plantation development is equivalent to impoverishment, and should have been anticipated in the planning stages of the resettlement program. This failure was acknowledged by the local government only in 2001, seven years after the households relocated from their original villages to Tanjung Balik and Tanjung Pauh.

Table 3.5 shows the education level of the poor households under consideration. The majority have only a low level of education. The poor of West Sumatra have more chances of obtaining higher education than those of the Koto Panjang resettlement villages. Poverty is also strongly associated with a low level of education. In West Sumatra, nearly 84 percent of poor households have been educated only to the elementary school level or below. In the Koto Panjang resettlement villages, more than 90 percent of poor households had an elementary school education or less.

Table 3.6 shows the size distribution of poor households. Large household size is not clearly associated with a higher poverty rate. Presumably as a result of Indonesia's household planning program, small household size is

Table 3.4 Main occupation of poor households (percent)

Main occupation	Resettlement village		West Sumatra
	Tanjung Balik	*Tanjung Pauh*	
Unemployed	22.9	15.0	12.9
Rice agriculture	34.4	12.5	54.8
Plantation	26.7	43.8	7.3
Animal husbandry	0	0	0.2
Fishery	3.1	1.3	2.6
Industry	0	0	0.7
Trade	2.3	1.3	2.3
Transport	0.8	0	1.3
Services	3.1	10.0	8.1
Others	6.9	16.3	9.8
Total	100.2	100.2	100.0

Note: Totals do not always add to 100 percent because of rounding.

Source: BPS (2006).

Table 3.5 Highest level of education of head of household (percent)

Level of education completed	Resettlement village		West Sumatra
	Tanjung Balik	Tanjung Pauh	
Elementary school or below	94.7	91.3	83.9
Junior high school	3.1	6.3	11.7
Senior high school or above	2.3	2.5	4.4
Total	100.1	100.1	100.0

Note: Totals do not always add to 100 percent because of rounding.

Source: BPS (2006).

a general norm. Nevertheless, having a smaller household apparently does not reduce the chance of living in poverty.

Resettlement villages in the Koto Panjang Dam project are not free from the risk of impoverishment. The authors concluded that failure in agricultural development is the main cause of poverty in Indonesia. Agricultural development has been a priority for the past three decades. In the Koto Panjang resettlement villages, development failure of the central government in agriculture has caused a long delay in terms of securing productivity of rubber plantations. The local government had to undertake costly development of rubber plantations after the resettlers arrived.

Problems and mitigation measures

The post-project survey carried out by JBIC (2004) revealed that the majority of resettled households in each West Sumatra resettlement village faced difficulties securing livelihood, mainly due to the failure in development of rubber plantations. While resettlers were supposed to receive productive rubber trees in their resettlement locations, few of the trees proved to be productive. In addition, the facilities for drinking water were not functional when people originally moved to these new villages in 1993.

Table 3.6 Size of poor households (percent)

Household size (persons)	Resettlement village		West Sumatra
	Tanjung Balik	Tanjung Pauh	
Fewer than 5	84.7	81.3	76.8
Between 5 to 9	14.5	18.7	21.6
More than 9	0.8	–	1.6
Total	100	100	100

Source: BPS (2006).

The government acknowledged the presence of several post-project problems. In 2002, the government formulated a Consolidated Action Plan (CAP) to be implemented between 2003 and 2010. The CAP included rehabilitation of rubber plantations through replanting, and construction of drinking-water facilities, bridges, and roads. Between 2003 and 2005, the government rehabilitated 1,600 ha of rubber farms (700 ha in 2003, 500 ha in 2004, and 400 ha in 2005). Of this, 900 ha of rubber farms were replanted for 450 resettled households in Tanjung Balik and 350 resettled households in Tanjung Pauh. Construction of the drinking-water facilities, bridges, and roads was completed in the same period.

The social impacts of the project were significant. According to the JBIC study, in both Tanjung Pauh and Tanjung Balik the majority of resettled households responded that their living conditions had worsened after relocation. Even after adjusting for the possible bias in the JBIC study, as mentioned earlier, it can be concluded that the resettlement of the two villages was not conducted in a satisfactory manner. The high incidence of poverty in the villages supports this conclusion.

The risk of impoverishment in the Koto Panjang resettlement villages can be attributed to insufficient implementation of the development plan. A failure in rubber plantations was a major factor disrupting the engine of economic life for the people of the Koto Panjang area. The introduction of suitable measures to ensure implementation of the plan could have avoided some risks. Moreover, the Koto Panjang resettlement program neglected to integrate and involve the stakeholders, such as local communities, local economic actors, local government, and central government agencies other than the PLN, the state power company. The lack of participation of local communities in the decision-making was one of the factors responsible for non-compliance with the development plan. The government seemed to acknowledge these sources of failure in the Koto Panjang resettlement program, but it has not yet reformed the development paradigm to enable greater integration of stakeholders and local economic development planning. The CAP should be replaced with a more comprehensive local development plan to reactivate local potential.

Nam Theun 2 (Lao PDR)

The Nam Theun 2 (NT2) hydropower project was built along the Nam Theun River, a tributary of the Mekong River in the central part of the Lao PDR. Key features of the project included constructing a dam on the Nam Theun River that would create a 450 km^2 reservoir in the Nakai Plateau, with a catchment area of 4,013 km^2.

According to NT2 Power Company, a total of 6,738 people in 1,298 households were impacted by the project. Among these, some 970 households (those that felt full impacts of the project) were fully eligible for the housing and livelihood rehabilitation program, while 94 households were

eligible for housing only, and 130 other households were eligible for livelihood programs only (partly affected households) (Nam Theun 2 Project
2004). Ethnicity was represented by six main ethno-linguistic groups: Brou
(40 percent of 1,298 households); Tai Bo (40 percent); Upland Tai (11 percent); Vietic (6 percent); Lao (2 percent); and Sek (1 percent). However,
distinctions between groups are blurred.

The NT2 Resettlement Action Plan has been heralded as a model for
future large hydropower projects involving resettlements in the Lao PDR.
The compensation policy of the NT2 project did not involve cash as a form
of compensation. Instead, the project provided livelihood facilities and
other infrastructure development (e.g., housing, water supply, agriculture,
public health, a village school, a village hall, village roads, and proximity to
the main road). The project did allow for some cash compensation where
there would be losses in agricultural production (such as fruit trees).

The project provided land for agricultural use in the resettlement area.
Before resettlement, traditional irrigated paddy rice fields amounted to
about 0.5–1.5 ha of land per household, and the upland rice fields were
about 1–2 ha per household. In the new resettlement villages land availability is limited, and the project was able to provide only 0.66 ha of farmland
per household. It turned out that the resettlers would have sufficient rice
only for about two to three months of annual household consumption.
Therefore, the project provided all the resettled communities with a total
forest area of about 18,207 ha (each of the 17 resettlement villages received
an average of 1,071 ha), which residents could use for materials to repair
their houses, to gather non-timber forest products, and for other daily uses.
They could also sell timber from the forests to generate household income.

The project built one house for each affected household of up to nine
members. Households of more than nine people were eligible to receive two
new homes. The project built new homes for all affected households. This
satisfied the resettlers' wishes, because new homes in the resettlement villages were of superior quality to those in the old villages.

A survey of four resettlement villages

The NT2 project affected 17 villages. For our research we conducted household
interviews in the four resettled villages of Boua Ma, Ca Oy, Done, and Sop On.
Field surveys were conducted December 15–25, 2010, and the sample included
135 households from the former four villages (50 households that now live
in Boua Ma, 35 in Ca Oy, 20 in Done, and 30 in Sop On). The villages were
resettled as follows (Figure 3.1):

1   The old village of Boua Ma was resettled as Boua Ma (name not
      changed) in 2006. Prior to resettlement, this village consisted of 76
      households and 369 people. At the time of resettlement, it received

4 households from other villages, bringing it to 80 households and 386 people. Two ethnic minority groups live in this village: there are 31 Brou households and 49 Makong households.

2   The old village of Ca Oy consisted of 35 households and 180 people. At the time of resettlement in 2005, these people were placed into two different resettlement villages: Sop On (31 households) and Done (4 households). The old Ca Oy contained two ethnic minority groups: 22 Tai Bo households and 13 Makong households.

3   The old village of Done was resettled as Done (name not changed) in 2005. Prior to resettlement, this village consisted of 145 households and 798 people. At the time of resettlement, it received 4 households from the Ca Oy resettlement, bringing it to 149 households and 812 people. Three ethnic minority groups live in this village: 4 Lao households, 1 Tai Bo household, and 144 Makong households.

4   The old village of Sop On was resettled as Sop On (name not changed) in 2005. Prior to resettlement, this village consisted of 105 households and 453 people. At the time of resettlement, this village received 31 households from the Ca Oy resettlement and 4 households from other villages, bringing it to 140 households and 633 people. Three ethnic minority groups live in this village: 111 Tai Bo households, 3 Lao households, and 26 Makong households.

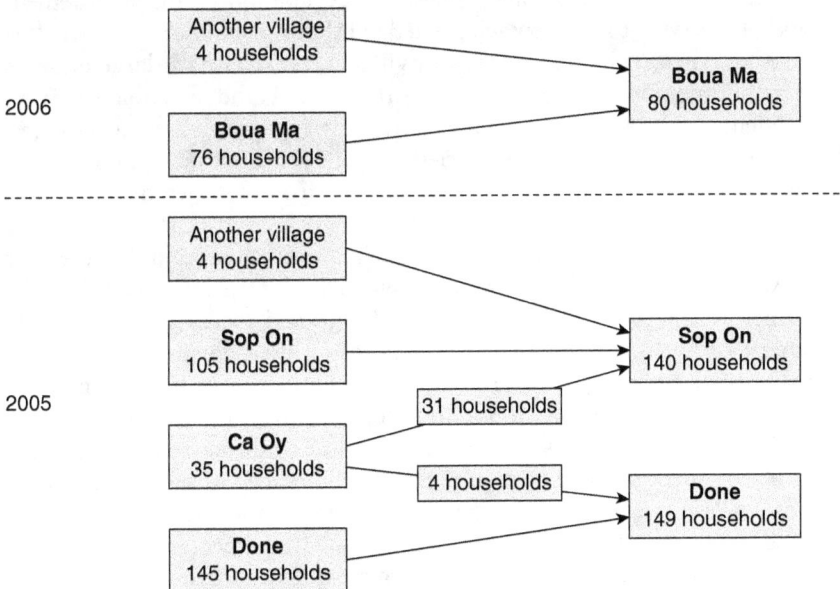

Figure 3.1 Resettlement of the four surveyed villages.

Income generation

Before relocation, the four surveyed villages (in their original locations) were fairly homogeneous in terms of occupations; most people were self-employed farmers, share-croppers, or laborers. In the old villages, there were no public or private sectors, although some young people in Boua Ma and Done worked as laborers for logging companies. In the resettlement villages, many of the occupations have been subject to change as the infra-structure and village facilities have changed. Most of the new resettlement villages are located near a main road and so some residents can work within the private and public sectors; however, many remain either self-employed farmers cultivating paddy rice or working as share-croppers.

In the old villages, it was a normal occurrence for residents to own buf-falos, pigs, and cows. These animals were maintained largely as capital assets for use in times of need. Thus, they were not considered a resource that would contribute to regular household income. A small proportion of the adult animals were used to plough in the field in those villages that had permanent rice fields. The value of one full-grown buffalo was about USD 200 before the time of resettlement. Thus, the main income in the old villages came from livestock. Almost every household raised poultry and pigs for household consumption and sometimes for ritual needs. After resettlement, residents continued the practice of animal husbandry. The Resettlement Management Unit of the NT2 Power Company also found this activity to be the main household income-generating occupation of the resettlement villages. However, it has been acknowledged that the number of livestock is decreasing when compared to the old villages, especially for large animals such as buffalo and cows, as a result of the limited land available.

Residents of the resettlement villages actively engaged in fishing activities. Fishing in the old villages was carried out in several natural ponds. All of these lakes have a high gradient that results from a series of waterfalls, rapids, riffles, and fast runs over stony and rocky substrate, occasionally with sand banks. In the new villages, residents can fish around the reservoir area, which is nearby. The reservoir is wider and deeper compared to the lakes before inundation. The average fishing yield increased significantly (from 19 kg to 40 kg annually per capita) after resettlement, thanks to the change of fish productivity, so resettlers could sell more fish to the market.

A comparison of average household income in the old villages versus the new resettlement villages is valid, since the value of the currency before and after resettlement did not change much. The average household income in the old villages was smaller, when compared to the average household income in the resettlement villages. On the one hand, people in the old vil-lages depended heavily on natural resources, since there were very few markets and it was difficult to access the town given the distance and the quality of the road. On the other hand, the current resettlement area has a market located within each village. The surveys conducted in December 2010 have shown that average yearly income is approximately USD 1,319 per household in Boua Ma, USD 1,192 in Ca Oy, USD 1,249 in Done, and USD 1,237 in Sop On. The income associated with all of these resettlement villages was higher than the Lao PDR rural poverty line (which was USD 850 per household per annum in 2009).

Merger of villages

During the resettlement, Ca Oy village was divided and merged into two other villages: Sop On and Done. Ca Oy included 35 households and 180 people. Of these, 31 households resettled in the new Sop On while four households resettled in Done. The villagers generally recognized that the merger was planned by the project. On the one hand, a majority of the villagers (80 percent and 77 percent in Done and Sop On, respectively) did not mind being merged with Ca Oy villagers. However, when they were asked whether they wanted to be resettled only with those from Done or Sop On, the majority of them (80 percent and 90 percent of Done and Sop On, respectively) answered yes. About half of the villagers (50 percent and 54 percent of Done and Sop On, respectively) felt that they experienced difficulties with the people from Ca Oy.

On the other hand, only 49 percent of Ca Oy villagers accepted the merger with other villages after relocation, and 37 percent did not. Among these villagers, 69 percent wanted to be resettled only with other Ca Oy villagers. About one-third (34 percent) of the villagers felt that they experienced difficulties with the people from Done or Sop On. None of the Ca Oy villagers indicated that after resettlement (within 4–5 years) their household members had married anyone belonging to another ethnic group.

Necessity of further development

The resettlers decided to live around the reservoir areas, and all indicated that they will continue to live in these resettlement areas despite limited sustainability of livelihoods. An expansion of the land used for agriculture is seen as one possible alternative for regional development.

About 18,206 ha (88 percent) of the land surrounding the reservoir is forested. The community forest area is sufficient enough to rezone some areas to serve as agricultural land for the resettlers. The NT2 project originally offered land for agricultural use at an average of 0.66 ha per household. In order to produce sufficient rice for year-round consumption the resettlers requested to be provided further land for shifting cultivation (about 1 ha per household, or about 1,298 ha for all resettlement households), in combination with the present land area of 0.66 ha, for a total of 1.66 ha per household (based on six persons per household).

With an agricultural area of 1.66 ha, the resettlers could obtain rice production of at least 2,480 kg based on a productivity of 1,500 kg/ha for rain-fed paddy fields of one crop per year. To ensure the success of the rezoning and resource use for resettlers and to improve the livelihood conditions of the resettlers, the Resettlement Management Unit of the project (in collaboration with resettlers) needed to discuss the following land- and resource-use management issues as of 2012:

1   consultative village land-use planning;
2   consultative allocation planning of the community forest area;
3   rain-fed paddy and arable fields;
4   non-timber forest products;
5   wildlife conservation; and
6   forest protection.

The difficulties observed in livelihood rehabilitation of the resettlers largely stem from the fact that resettlers were obliged to convert from a mixture of nomadic and slash-and-burn farming to intensive agriculture with small farmlands. Moreover, water was not constantly available for intensive rice cropping in the newly developed farmlands for resettlers. Their production base was thus largely degraded after relocation. An officer of the NT2 hydroelectric project responsible for the resettlement scheme admitted that the existing production base was not sufficient for the resettlers to maintain their livelihood and that some corrective measures were needed, such as giving the resettlers permission to use the public forests for production purposes (personal communication, February 13, 2012).

The authors regard the concession period of 31 years for the entire dam project (Nam Theun 2 Project 2005), of which the operating period is 25 years, as the major source of the observed problems. All the systems needed to be built in six years to make the hydropower station operational for 25 years. The area for resettlement was thus confined to a small, newly developed area on the Nakai Plateau that was not appropriate for nomadism or slash-and-burn farming. Finding a larger area, to have the resettlers maintain their traditional occupation of nomadiss and slash-and-burn farming, was not considered in the development of the resettlement scheme because of the paucity of time for planning.

Kotmale Dam (Sri Lanka)

Mahaweli Development Program and Kotmale Dam

The Kotmale project is one of five major headworks projects (a civil engineering term for any structure at the head or diversion point of a waterway) that were undertaken under the Mahaweli Development Program. Financially assisted by the government of Sweden, Kotmale Dam is the furthest upstream of the projects. It was developed to regulate river flows in addition to harnessing the hydropower potential of a major right-bank (when facing downstream) tributary of the Mahaweli River—the Kotmale Oya, which flows through the rural mountain regions of Sri Lanka, passing ancient villages steeped in history and tea plantations of a more recent era.

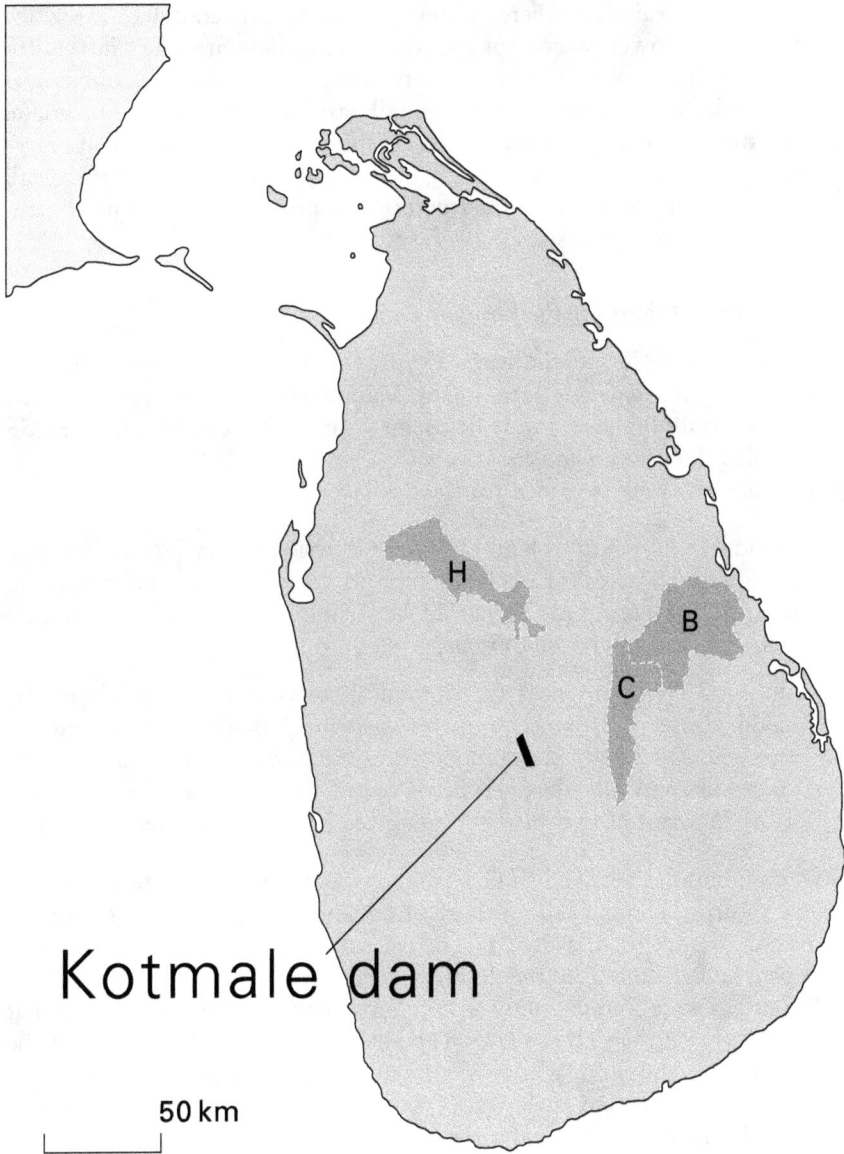

Figure 3.2 Location of resettlement sites in Sri Lanka.

Source: Planning & Monitoring Unit, Mahaweli Authority of Sri Lanka.

The Mahaweli Development Program covers approximately 210,000 ha of farmland, where nearly a million persons were settled during the 1970s and 1980s, including resettlers from the four dams (Victoria, Randenigala, Rantembe, and Kotmale) constructed under the project (see Box 3.1 at the end of this chapter). The command area has been divided into "Mahaweli systems" (sub-regions), where Systems A to G are contiguous regions located on the lower reaches of the Mahaweli. These areas are in the dry zone, which covers almost three-quarters of the country and is characterized by a long dry season, high annual rainfall variability, and a warm climate. To improve the quality of life of resettlers (both voluntary and involuntary) of the Mahaweli Development Program, a well-planned physical, social, institutional, and economic infrastructure was provided in these newly created resettlement schemes (Mahaweli systems).

Resettlement options for the Kotmale Dam

From the late 1970s to the early 1980s, the Kotmale Reservoir flooded nearly 4,000 ha of fertile land in the Mahaweli upper catchment, which included about 600 ha of paddy fields, and caused the resettlement of 3,056 households due to inundation.

Two options were provided for the displaced:

> Option 1: Agricultural land (2.5 acres of irrigated dry land and/or rice fields and 0.5 acres for the home plot) in the new Mahaweli Systems B, C, and H (Figure 3.2). The 1,722 households that selected this option were allowed to choose to which system to move.

> Option 2: Tea plots near the Kotmale Reservoir, a scheme selected by 1,334 households, which were resettled in 17 settlements around the reservoir. The size of the compensation tea plots was determined based on productivity: 0.51 ha (1.25 acres) of low-producing seedling tea or 0.3 ha (0.75 acres) of vegetatively propagated higher-yielding tea allotments.

Both communities received similar compensation packages in terms of economic returns. Though falling short of best-practice guidelines, the provisions did attempt to alleviate some of the hardships of relocation to ensure that people benefited from the development.

Resettlers were initially allowed to choose between the two options but could not revisit their choice later. The authorities delayed handing out the legal titles for the land, which was intended to prevent speculators from prompting resettlers to make hasty sales and depart. Substantial investments were made in technical assistance and training in agriculture-related activities, with the objective of protecting such new land titles at least until the resettlers had established new production systems and could make informed judgments about likely earnings.

The study's approach

This study's findings are based on several field visits to the resettlement sites, in which interviews were held with households using a structured questionnaire.

Key informants and officials were interviewed separately. Two household surveys were conducted: one in 2005 (Takesada et al. 2008) and another in 2011. The surveys covered 266 and 171 households, respectively.

For any kind of post-project review, the assessment of social impacts can follow the "impoverishment model" as suggested by Cernea (2000). In this model there are eight forms of impoverishment: landlessness; joblessness; homelessness; marginalization; increased morbidity and mortality; food insecurity; loss of access to common property; and social disarticulation. These parameters can be used to identify the strengths of the resettlement scheme, to review and assess the negative impacts, and to propose measures for prevention or mitigation.

The questionnaire for the 2005 survey comprised the following elements:

1   Biographical and personal attributes of respondents.
2   Economic activities to measure levels of satisfaction; opportunities created or lost after resettlement.
3   Respondents' satisfaction with respect to physical or material well-being.
4   Health status of resettlers; improvement or deterioration of household sanitation, water supply, etc.; education and related issues.
5   Questions to measure self-assessment and perceptions regarding the resettlement scheme, including self-esteem; evaluation of new opportunities and general satisfaction; review of the compensation scheme to identify resettlers' views.

The questionnaire for the 2011 Survey, aimed at collecting further details, included:

1   Economic conditions (trends in the household economy); perceptions of the community.
2   Assessment of satisfaction regarding irrigation and water availability; roles of farmer organizations.
3   Income-generating activities (on-farm, off-farm, and non-farm).
4   Community participation and institutional support in rebuilding social infrastructure.
5   Improvement or impoverishment of quality of life, social problems, safety of people with respect to human–elephant conflicts, etc.

The two authors, together with four local field assistants, carried out the interviews, during which the first author interpreted and translated the questions and responses.

Household surveys that seek clues about satisfaction are difficult to administer or interpret because respondents often lack a context for or explanation of the questions; therefore, they often skip questions or provide vague responses. In the surveys presented here, attitudes were appraised indirectly, by asking respondents about their reactions to hypothetical situations—for instance, by asking them what they would opt for if they were given a choice of the place of resettlement now.

Data are scanty on the cropping patterns and incomes of prospective resettlers in the pre-settlement villages. General profiles for traditional rural households are available, but not for Kotmale. We asked the interviewed householders to compare present earnings from different sources with income levels before they resettled. Few quantitative references to pre-settlement net incomes refer to wage earnings as laborers or sharecroppers, or cash sales of crops. The omission of subsistence products consumed at home is noteworthy. Such goods, including firewood, grazing fields, etc., were the basis for the stable household economy many resettlers have lost. Reflections on pre-settlement lifestyles among those who are involuntarily moved always tend to exaggerate the agreeable conditions of yesteryear (World Bank 1998). Even so, comparisons of pre- and post-dam cash incomes do not reflect the benefits of the earlier lifestyle that now matter most to the resettlers (World Bank 1998).

Repeated meetings with a small number of households, supplemented by interviews with key informants and community groups, are an effective, low-cost technique for tracking the performance of rural development projects (World Bank 2004) such as the Kotmale case study presented in this chapter.

Observations

General levels of satisfaction

People who were resettled in the vicinity of the reservoir, i.e., the Kotmale area, were more satisfied than those who had moved to Mahaweli Systems B, C, and H. (Results of the questionnaire survey are shown in Table 3.7.)

More than 95 percent of those who settled in Kotmale were satisfied with the resettlement option, whereas rates among those who had moved to

Table 3.7 Satisfaction with choice of resettlement

Are you satisfied with the choice you made in selecting the resettlement option?	Kotmale		System B		System C		System H	
	House-holds	%	House-holds	%	House-holds	%	House-holds	%
Yes	67	95.7	47	73.4	38	70.4	56	71.8
No	3	4.3	17	26.6	16	29.6	22	28.2
Total	70	100	64	100	54	100	78	100

Source: Authors.

Table 3.8 Choice of resettlers, given the choice of selection now

If you were given the chance, would you select the same choice now?	Number of households			
	Kotmale	System B	System C	System H
Yes	69 (67)	45 (47)	34 (38)	56 (56)
No	1 (3)	19 (17)	20 (16)	22 (22)
Total	70	64	54	78

Note: Numbers in parentheses represent the number of households who expressed their satisfaction for selection of the particular resettlement option given in Table 3.7.

Source: Authors.

Mahaweli systems were around 70 percent. A small proportion of people, especially in Systems C and H, replied that they would be satisfied provided that certain conditions were fulfilled, such as the availability of more irrigation water and more options for the second generation of resettlers.

To check whether the responses of the resettlers were consistent and reliable, we asked the following hypothetical question (at the end of the questionnaire): "What would your choice be if you were given the chance to select the place of resettlement now?" The responses obtained are summarized in Table 3.8.

The number of households that were satisfied with the choice of resettlement (as indicated in Table 3.7) closely matches the responses given for the choice of resettlement if made now (Table 3.8). This is an indication of the dependability of the responses given by the households.

Reasons for selecting the resettlement scheme

The majority of the households—about 80 percent of those who moved to System C, and more than half of those who moved to the other two systems— wanted to move to Mahaweli systems because they wanted to continue paddy cultivation (Table 3.9). The second noteworthy reason was the influence of relatives who had decided to move to these areas, which indicates that some households made collective decisions to move to the same area.

The numbers in parentheses in Table 3.9 represent the number of households that expressed satisfaction in their selection of a particular resettlement option, results which were obtained from the questionnaire survey. It is clear that the majority of those who chose paddy cultivation as the reason for choosing the resettlement option were generally satisfied. For example, 35 of 37, 32 of 43, and 39 of 41 households that decided to move to Systems B, C, and H, respectively, in order to continue paddy cultivation were satisfied with their choice of resettlement. In contrast, the number of satisfied households was smaller for those who moved for other reasons.

Table 3.9 Reasons for selecting the resettlement option for those who moved to Mahaweli systems

Reason for choosing the resettlement option	System B		System C		System H	
	Households	%	Households	%	Households	%
Like to cultivate rice paddy	37 (35)	57.8	43 (32)	79.6	41 (39)	52.5
Received more land	2 (2)	3.1	1 (1)	1.9	3 (1)	3.8
Relatives moved	20 (9)	31.2	7 (3)	12.9	14 (6)	17.9
Did not like to grow tea	0	0	3 (2)	5.6	9 (6)	11.5
Had no option	4 (0)	6.3	0	0	5 (1)	6.4
Other reasons	1 (1)	1.6	0	0	6 (3)	7.6
Total	64	100	54	100	78	99.7

Note: Regarding parenthetical numbers (number of households who expressed satisfaction in their selection of a particular resettlement option), see Table 3.8. Totals do not always add to 100 percent because of rounding.

Source: Authors.

Those who wished to settle near the reservoir decided to do so because they did not want to move to Mahaweli systems due to the harsh climate of the dry zone of Sri Lanka (Table 3.10). Also, they wanted to continue to live in Kotmale for reasons such as a wish to live in their ancestral villages, and had less willingness to leave traditional villages whose locations are conveniently located close to established urban centers.

Traditionally, farmers from the Kotmale area relied on well-adapted wet-rice cultivation along river valleys. Those resettlers who decided to stay in Kotmale faced a problem in being relocated high up in the hills. Given the total change of terrain and the insufficient water, they had to abandon wet-rice cultivation and start as small-scale tea cultivators. In fact, their new land plots were subdivisions of unproductive tea estates.

Table 3.10 Reasons for selecting the resettlement option (vicinity of the reservoir)

Reason for choosing the resettlement option	Households	%
Had knowledge about tea cultivation	9	12.8
No household labor	6	8.6
Climate	45	64.3
Wanted to be at Kotmale	10	14.3
Total	70	100

Source: Authors.

Agricultural extension services to care for this new group of tea growers, and especially their need for training, seems to have been almost non-existent. Resettlers in the Mahaweli systems, although they were able to continue growing rice, had to cope with a new situation, implying increased market integration and commercialization of the whole agricultural sector. This, in effect, meant that they found themselves changed overnight from small-scale, mixed-cropping, subsistence-oriented peasants to farmers producing a cash crop based on capital-intensive technology (Søftestad 1991).

Reasons for satisfaction or dissatisfaction

Level and stability of income

The levels of income since resettlement increased for nearly 60 percent of the resettlers in all four resettlement areas (Kotmale and Systems B, C, and H) (Table 3.11). However, the stability of income decreased after resettlement by more than 50 percent for those who moved to Mahaweli systems, whereas it increased by 63 percent of those who resettled near Kotmale.

The satisfaction with the choice of resettlement has a direct relation with the increase of income levels. Almost all households whose income increased after resettlement were satisfied with their choice of resettlement, as seen by the numbers in parentheses (Table 3.11). This observation is also true in

Table 3.11 Change in levels of income and stability of income, by resettlement area

	Kotmale		*System B*		*System C*		*System H*	
	Households	*%*	*Households*	*%*	*Households*	*%*	*Households*	*%*
Level of income								
Increased	41 (41)	59	38 (35)	59	31 (29)	57	47 (44)	60
Decreased	7 (4)	10	10 (2)	16	9 (4)	24	15 (2)	19
No difference	22 (22)	31	16 (10)	25	14 (5)	19	16 (10)	21
Total	70 (67)	100	64 (47)	100	54 (38)	100	78 (56)	100
Stability of income								
Increased	44 (44)	63	31 (30)	33	19 (19)	35	20 (20)	26
Decreased	19 (16)	27	25 (10)	53	29 (15)	54	39 (22)	50
No difference	7 (7)	10	8 (7)	14	6 (4)	11	19 (14)	24
Total	70 (67)	100	64 (47)	100	54 (38)	100	78 (56)	100

Note: Regarding parenthetical numbers, see Table 3.8.

Source: Authors.

connection with stability of income. When the stability of income increased, the resettlers were more likely to be satisfied with their choice of resettlement, which was one of the reasons why those who resettled near Kotmale were more satisfied than those who moved to Mahaweli systems. During informal interviews, the key informants reiterated this: more stable income was as important as increased levels of income. For those who resettled near Kotmale, both the levels and stability of income increased, leading to relatively higher levels of satisfaction.

Land ownership

The extent of land ownership increased by more than 60 percent for all resettlers (Table 3.12). This again corresponded well with their level of satisfaction. The households that had a smaller land extent after resettlement, especially among the households that moved to System H, were not satisfied with their choice, mainly because of the decreased land ownership.

Availability of irrigation water

The majority of resettlers were satisfied with the irrigation water availability, which included attributes such as quantity available, timely supply of water, and overall management of the system by farmer organizations (Table 3.13).

The majority of resettlers were satisfied with irrigation water availability at the time of the survey, though the majority had been unsatisfied soon after resettlement. Subsequent developments in irrigation water management, especially the input by farmer organizations, were crucial in shaping the levels of satisfaction of resettlers. The increased availability of irrigation water will lead to successful paddy cultivation, together with higher income patterns, which will ultimately lead to overall satisfaction of the resettlers. In contrast, one area with which most respondents were dissatisfied was

Table 3.12 Change of land extent among resettlers

	Kotmale		System B		System C		System H	
	Households	%	Households	%	Households	%	Households	%
Increased	45 (42)	64	42 (37)	66	37 (32)	69	48 (43)	62
Decreased	20 (20)	29	15 (7)	23	12 (4)	22	25 (9)	32
No difference	5 (5)	7	7 (3)	11	5 (2)	9	5 (4)	6
Total	70 (67)	100	64 (47)	100	54 (38)	100	78 (56)	100

Note: Regarding parenthetical numbers, see Table 3.8.

Source: Authors.

Table 3.13 Satisfaction with irrigation water availability (percent)

How satisfied are you with availability of water in your farmland?	System B (N = 42)		System C (N = 49)		System H (N = 46)	
	Soon after resettlement	Now	Soon after resettlement	Now	Soon after resettlement	Now
Satisfied	54	78	45	81	61	85
Somewhat satisfied	15	13	12	5	6	8
Not satisfied	31	9	43	14	33	7

Source: Authors.

assistance with paddy rice marketing and the selling of their produce at reasonable prices, though every year the government offers a guaranteed minimum price to buy paddy rice.

Findings

Of particular interest in this case study is why more of the households that resettled in the vicinity of the reservoir (>95 percent) were satisfied with their resettlement arrangement, compared to those who moved to Mahaweli systems (around 70 percent), despite receiving similar compensation packages in terms of income generation. Moreover, land ownership increased for those who moved to Mahaweli systems, with new infrastructure facilities to rebuild the social framework.

Increased income patterns and land ownership were common factors for satisfaction; however, income instability thwarted this sense of improvement. Households that resettled in the vicinity of Kotmale were more satisfied, and therefore it is apparent that the satisfaction levels among resettlers largely depend on circumstances other than what have been stated above. Two comparisons can be made:

1   Compare the proportion of satisfied households between those who resettled in the vicinity of the reservoir (Option 1) and those who moved to Mahaweli systems (Option 2).
2   Compare the proportion of satisfied households between those that selected different locations under Option 2, whether Systems B, C, or H.

It was observed that education opportunities for the second generation were greater for Kotmale resettlers compared to those who moved to Mahaweli systems, thanks to better facilities for secondary and higher education. The percentage of persons completing education up to secondary levels increased after resettlement for all the locations (Takesada et al.

2008). Nearly 9 percent of those resettled in the vicinity of the reservoir and 3 percent of those who settled in System H had the opportunity to obtain higher education. (In contrast, no resettlers in Systems B or C had obtained higher education by the time of the survey.) The extra opportunity for their children to obtain higher education led to satisfaction with their locations of resettlement (Takesada et al. 2008). Kotmale and System H are located with easy access to major townships with well-equipped schools. It appears that, in moving to Systems B and C, the resettlers had to sacrifice—probably unknowingly—educational opportunities for their children. Another reason that can be highlighted is the priority given to land ownership over other material benefits, which can be related to levels of poverty, landlessness, and unavailability of information as to how they should plan their future lives. Without production assets to pass on to dependants, parents were left with little choice in supporting the second generation, as described in detail by Takesada et al. (2008) and Manatunge et al. (2009).

In addition to the abovementioned observations, the following are other reasons identified from interviews as to why the resettlers who moved to Mahaweli systems were not satisfied with their choice of resettlement location:

1   Harsh climatic conditions in the dry zone, compared to the mild conditions prevalent in Kotmale.
2   Increased incidence of diseases, such as malaria, kidney disease, and high blood pressure (ascertained only from household responses), which led to impoverishment, both economically and socially.
3   Human–elephant conflict, which led to much destruction, severe stress among farmers, and wasted time and money in protecting their farmland and produce.

Attitudes differ, and there is no consensus between resettlers as to what leads to satisfaction, such as better productive farmland with irrigation water and fertilizer subsidy, more accessible schools and health facilities, off-farm income sources, etc. Judging by the actions of the households and their answers during interviews and other discussions, it is clear that some resettlers were satisfied despite insecure and unstable levels of outcome. Conversely, a small number of respondents said that their livelihood conditions and prospects for better economic outlook had deteriorated since resettlement. Yet, with some exceptions, they agreed that their homes were better built and farms better irrigated since resettlement.

The observations of this study show that negative consequences of resettlement are often offset by favorable factors such as access to irrigation infrastructure and other institutional support. This follows a similar observation to that reported for aquaculture development, for which proper management of aquatic resources is the key to maximization of benefits

(Manatunge et al. 2009). Some farmers who had lost land to resettlement had valid grounds to reject the favorable comparisons between their present and past conditions that show satisfactory and stable incomes after resettlement. They claimed that although the whole Mahaweli program remained stable, they were not happy, because their income levels and stability of income had both decreased. In addition, harsh climatic conditions to which the resettlers are averse had led to more frequent health-related problems, thus affecting their physical capability, which determines whether they are fit and capable enough to engage in farming, share-cropping, or working as laborers. Most households that did not have household labor had to depend on hired labor, which had led to deterioration of their profits, thereby rendering their farming activities unprofitable (Takesada et al. 2008; Manatunge et al. 2009).

The primary rationale for the Mahaweli Development Program—rice import substitution—ceased to be tenable shortly after the resettlement program was commissioned, back in the early 1980s, because of a sharp drop in world rice prices (World Bank 2004). This unquestionably led to decreased economic viability for rice cultivation, which had been the only source of income for most of the resettlers who moved to Mahaweli systems. Diversifying out of paddy farming was not realized for so long because the policy environment and the mind-set of Mahaweli officials were not conducive at that time, a circumstance that remains largely true even today. The resettlers who moved to Mahaweli systems were accustomed to deriving income from off-farm sources (e.g., rental of equipment, hiring of labor) and non-farm sources (wage work, cottage industries, employment outside or abroad) before they were resettled. However, with resettlement they lost such additional income sources, until re-establishing such activities a decade or so later.

Resettlers from Kotmale were not accustomed to large-scale paddy cultivation. Most of them did not have such farm skills as irrigation water management and crop selection depending on water availability. Also, Mahaweli staff members were not trained to foster transfer of such skills. There was a strong tradition from the time of resettlement to focus on overall paddy production targets rather than market-driven demand for produce. These circumstances always led to gluts and lower market prices, which resulted in less stable income patterns and therefore dissatisfaction among resettlers.

Although the majority of resettlers decided to move to Mahaweli systems due to their intention to do paddy cultivation, the resettlers were not provided with full title to their land until after a considerable delay, which is one of the reasons they could not prosper. The title was not given because of a paternalistic concern that they would speculate with the land rather than cultivate it. However, this was counterproductive; farmers did not have any other property for obtaining loans to offer as collateral, which was a reason most of the resettlers were unsatisfied.

Irrigation water needs to be priced to reflect the sectoral need for financing to meet recurrent expenditure and capital recovery. Irrigation operation and maintenance costs need to be recovered from users if such large-scale projects are not to be an excessive burden on public finances. Mahaweli systems would have catered better to the needs of the resettlers had such a cost-recovery system been in place.

Resettlers who settled in Kotmale after receiving tea plots did not face this scenario of declining prices for their produce. Their main difficulty was the unproductive tea plots they received and lack of capital for investing in replanting tea or purchasing regular stocks of fertilizer to enhance the productivity of their limited land holdings (Manatunge et al. 2009). Manure and inorganic fertilizers are relatively expensive, and present yields of tea plots and incomes are too low to justify large expenditures on these purchased inputs. For many farmers, it was impossible to escape from this low-productivity trap.

Mahaweli irrigation resettlement schemes were well-connected to other parts of the country with a network of well-planned roads. Access to these schemes has been continually improved and has recently been upgraded with the establishment of new city centers, providing further opportunities for enhanced marketability of surplus crops.

A majority of the Kotmale resettlers were able to overcome the difficulties they faced at the time of resettlement, and managed to restore their livelihoods. Such success stories are available elsewhere in Sri Lanka; the most commonly cited examples are the Pimburettewa Scheme (1971) and the Victoria Dam Project (1984). But it is difficult to assess how resettlers were affected in most of the cases because there are no or little data available on the situation prior to evacuation.

More farmland and increased income were some of the factors that contributed to satisfaction among the farmers who moved to Mahaweli systems. It appears that those resettled in the vicinity of the reservoir were more satisfied than those in Mahaweli systems, mainly thanks to more opportunities for the second generation (e.g., better opportunities for education) and higher income stability. Thus, it is important to give priority to the needs of second and future generations for sustainability of any resettlement scheme.

A majority of the households that moved to Mahaweli systems wanted to continue paddy cultivation as their main source of income. Land extent and water availability were satisfactory, but the income of resettlers was not stable due to market price fluctuations that made them develop negative perceptions about resettlement. This shows that satisfaction in the choice of resettlement has a direct relation to stability of income. Another factor in the resettlers' satisfaction was better irrigation infrastructure and water availability, which included attributes such as quantity available, timely supply of water, and overall management of the system by farmer organizations. This implies that land-based compensation should always be supplemented by appropriate irrigation infrastructure, which will assure success of the resettlement scheme. Other factors such as harsh climatic conditions, increased incidence of diseases, and conflicts with wildlife also led to difficulties after resettlement, which caused much discomfort among resettlers. Success of any resettlement scheme depends on such secondary factors too; they should be eliminated to the extent possible, as they will indirectly lead to lower quality of life, including economic hardship.

Diversifying out of paddy cultivation during the periods when rice prices continued to fall would have made the resettlers gain more from resettlement compensation. Therefore, diversification and introduction of alternative sources of income, sustainability of production capacity, and economic viability in the long term are essential considerations in resettlement planning that may guarantee the establishment of livelihoods and stability of income.

Awarding land titles on a timely basis would have provided opportunities for resettlers to secure loans. However, awarding such legal title would also have led to land speculation, which can be counterproductive and leave the resettlers landless. In any case, livelihood-rebuilding efforts should be complemented with opportunities for securing financial assistance and access to credit, which is crucial in the success of any resettlement scheme.

Sameura Dam (Japan)

The Shikoku-Region Development Promotion Act was put in place in 1960 for the purpose of developing natural resources in Shikoku, the smallest and least populated of the four major islands of Japan. Among the projects associated with this initiative, the water-resource development plans for the Yoshino River, a major river in the Shikoku region, were regarded as core projects. The Sameura Dam was built in 1973 as a part of the water-resource development plans for the Yoshino River. It was intended to supply large amounts of water to the neighboring prefectures, and its construction made a significant contribution to the economic development of the Shikoku region.

The areas affected by this project include the towns of Motoyama and Tosa, and the village of Okawa. Among these, Tosa and Okawa were the main areas inundated by the dam reservoir. The resettled individuals from this area included 352 households (Mizushigen Kaihatsu Koudan 1979). Although the submerged area of Okawa was smaller than that of Tosa, a central portion of Okawa, including government offices, schools, and businesses, was submerged. Therefore, Okawa was most significantly affected by the construction; 141 out of 167 households moved to neighboring towns or other cities, causing a sharp decrease in the population (Okawa Mura 1981). Furthermore, the mining industry was the main industry in the village and included over 2,000 workers. This industry was forced to shut down in 1971 due to trade liberalization that emerged after construction began. This further accelerated the exodus from the village, and the population decreased from 3,212 to 2,206 at that time (Mizushigen Kaihatsu Koudan 1979). Those who remained in the village attempted to redevelop the region and industry by engaging other local resources (e.g., forests and livestock); however, these efforts were complicated by rapid depopulation, and, in some cases, abandoned.

Resettlement negotiations

The village of Okawa suffered the most significant destructive impacts from the dam project. In view of the anticipated impacts, the villagers had strongly protested against the project. In 1962, the villagers from Okawa and Tosa jointly passed a resolution in opposition to the dam construction and established the Alliance for Construction Resistance (Okawa Mura Shi Tsuiroku Hensan Iinkai 1984). In that same year, the village constructed a new public office within the projected dam site as a symbol of protest. The villagers also refused external investigation aimed at evaluating the worth of their properties and fields. The people of Okawa worked with a coalition to negotiate against the WRDPC and formed the new Council for Countermeasure for Resettlers in 1965.

After repeated requests over several years to halt construction, the village of Okawa finally agreed to have the survey conducted for the sake of dam construction in 1966. This agreement was reached because it had become difficult to consolidate the diverse opinions of the many people living in Okawa and other affected areas. Added pressure was placed on those living in Okawa because other affected areas, including the towns of Tosa and Motoyama, had agreed to the proposed survey in 1964. In addition, the government of Kochi Prefecture began to invest serious effort into methods that would secure provisions for those forced to resettle elsewhere. The prefectural government initiated promotion of a development plan in the area and began consultations with those residing in the affected area (Mizushigen Kaihatsu Koudan 1979). However, the prefectural Assembly responded to these actions with displeasure because the prefecture did not reveal any position regarding dam construction until July 1965. The preparatory work for dam construction began in 1967 and construction of the dam was completed in 1973.

Compensation schemes for the Sameura Dam resettlement

The compensation standards for the Sameura Dam construction were settled in 1967, after more than 100 negotiation meetings (Okawa Mura Shi Tsuiroku Hensan Iinkai 1984). With regard to individual compensation, cash compensation equivalent to the material loss was paid in accordance with the standards. A group of negotiators from each affected village and town was put in place, and each affected area sent an equal number of representatives to the negotiation table. One of the reasons that there was such a long negotiation period was the skyrocketing prices of land in the planned resettlement area adjacent to the dam site (i.e., the towns of Tosa and Motoyama). The proposed compensation standards could not cover the price of alternative land in these neighboring areas (Mizushigen Kaihatsu Koudan 1979). Many people in the affected areas would not agree to the proposed compensation, because they never could have found comparable land in the resettlement area for the price that they were being offered to give up their existing land. At the same time as the negotiations for individual compensation standards were taking place, the land expropriation for a new road within Motoyama town was settled, and this was considered a guide for compensation on the dam. As a result, negotiations for individual compensation resumed.

The negotiations for public compensation, including public facilities and residential complexes, began in 1967; this was after individual compensation negotiations were initiated in 1966 (Okawa Mura Shi Tsuiroku Hensan Iinkai 1984). The letter of agreement on public compensation (including the replacement of roads and the development of a new residential area) was finally signed in 1968.

At that time the villagers requested a special subsidy for community rehabilitation that would include development of an agricultural park and a lodge. But the village of Okawa could not attain the desired subsidy to rehabilitate their livelihood, because the dam construction began before the Act on Special Measures concerning Measures Related to Water Resource Areas (see Box 2.1) came into force in 1973. Consequently, resettled individuals received payment only for their material loss, and did not receive any additional compensation to secure future livelihood.

The process and length of compensation negotiations have an influence on long-term community retention. The Sameura project demonstrates some great lessons for all stakeholders concerning dam construction and how compensation negotiations ought to proceed. The negotiation of individual compensation that took place with the project prior to determining the public compensation package undermined the community bonds among villagers. The end result was that the village and villagers missed the opportunity to discuss their future due to rapid depopulation and the unexpected closure of the mine.

An uncertain redevelopment plan in terms of public compensation has a destructive impact on community rehabilitation in the long term. The Sameura Dam project showed that a vague future plan will discourage individuals from remaining in the village. The best future plan includes livelihood rehabilitation measures such as secure transportation, schools, job opportunities, and public services that will mitigate the emotional strain on resettled individuals.

Discussion

The problems that emerged from the five resettlement programs in four different countries presented in this chapter should have been avoided by proper planning, but could have also been solved even during the implementation process.

In the case of the Sameura Dam, the people of Okawa village—the community most adversely affected by the dam construction—strongly opposed the resettlement. When they finally and reluctantly agreed to resettlement, they lost not only an opportunity to negotiate with the dam developer regarding compensation for public facilities, but also an opportunity to establish a long-term livelihood rehabilitation plan. They were only provided with compensation for private property, as other affected communities had already agreed to resettlement and the land had already been transferred to the dam developer. Ideally, a comprehensive resettlement master plan should be developed through the complete participation of all affected communities. In this case, the dam developer should have made every effort to involve all affected communities. Even if some affected communities are strongly opposed, developers should continue working to encourage their participation in planning, by presenting information about the implementation of resettlement that has satisfied other affected communities. Even in such a case, late resettlers should be given fair and equitable treatment compared to early resettlers. Developers should provide compensation for public facilities even after communities complete their resettlement.

As for resettlement plans, fairly generous compensation was to be provided for resettlers of the Bili-Bili and Koto Panjang Dams in Indonesia. In both cases, submerged properties were compensated with cash. In addition, resettlers from the Bili-Bili Dam were given an opportunity to participate in government's TP, which was exclusively planned for transmigrants from Java and Bali to other less-populated islands. Resettlement to make way for the Koto Panjang Dam was conducted village by village, and farmland and houses were provided free of charge based on the same conditions as in the TP. If both resettlement programs had been properly implemented, resettlers would have been able to effectively rehabilitate their lives. Things did not turn out that way, however. Resettlers suffered hardships for some years after resettlement and their livelihood rehabilitation process took much longer than expected. At the resettlement areas of the Bili-Bili resettlers, infrastructure was not properly prepared, conflicts occurred between resettlers and local communities, and in some cases farmland provided was much smaller than promised. As a result, many of the resettlers returned to the reservoir area. When Koto Panjang Dam resettlers arrived at the resettlement areas, farmland development was not yet complete, so they lost their income sources. Dam developers might regard the development of resettlement areas to be less important than the actual dam construction, since the former may not generate economic benefits for them. Both are inseparable, however, so good systems to monitor progress in preparing resettlement areas are essential.

Similar lessons can be learned from Kotmale Dam resettlement in Sri Lanka. Resettlers were given an opportunity to participate in a national resettlement program. Development of irrigation facilities was not sufficient, however, though less serious than in the Indonesian cases. Problems could have been avoided by a greater emphasis over dam construction. Another lesson from the Kotmale case is that resettlers valued income stability more than an increase in income. In that case, the international price of rice had become unstable after resettlers had resettled, and this decreased their satisfaction with the arrangements. A counter to such an outcome would be to diversify resettlers' income sources, and dam developers could play a role here. Resettlement programs should be flexible enough to be modified so that resettlers can adapt to social and economic changes after resettlement. If educational opportunities decrease after resettlement, diversification of income sources may be more difficult. Options for education and employment for the second generation of resettlers should be considered through various measures, including new educational facilities and scholarships.

Farmland originally given to the NT2 hydropower project resettlers in Lao PDR was only 0.66 ha for each household. This was too small for them to secure the necessary food supply for themselves. Moreover, insufficient training on paddy cultivation was provided for resettlers who used to conduct slash-and-burn agriculture. While the dam developer granted resettlers additional access to utilize the community forest, it should have further implemented measures to diversify resettlers' income, such as through the promotion of a fishery at the reservoir. In Chapter 4 we will present a successful case of income increase and diversification of a community of NT2 resettlers by producing handicrafts.

We can extract two major lessons from the five cases of dam resettlement programs presented in this chapter. First, while every effort should be made to plan resettlement programs for the long term, dam proponents and developers should continue striving to provide resettlers with further options, whenever necessary, to choose income sources and jobs that will help them adapt to social and economic change. Second, the preparation of resettlement areas should be completed before people start to move; this preparation should be given a high priority, even over dam construction and inundation, and progress should be properly monitored.

References

BPS (Badan Pusat Statistik) (2006). *Selected Indicators of Socio-Economic Indonesia, July 2006 edn.* Jakarta: BPS.

Cernea, M. M. (2000). Risks, safeguards, and reconstruction: a model for population displacement and resettlement. *Economic and Political Weekly*, 41, 3659–3678.

JBIC (2004). *Kotapanjang Hydroelectric Power and Associated Transmission Line Project (1) (2), Third Party Ex-Post Evaluation Report.* Retrieved from www.jica .go.jp/english/our_work/evaluation/oda_loan/post/2004/pdf/2-06_full.pdf.

Karimi, S., Nakayama, M., Fujikura, R., Katsurai, T., Iwata, M., Mori, T., and Mizutani, K. (2005). Post-project review on a resettlement program of the Kotapanjang Dam project in Indonesia. *International Journal of Water Resources Development*, 21 (2), 371–384.

Manatunge, J., Takesada, N., Miyata, S., and Herath, L. I. (2009). Livelihood rebuilding of dam affected communities: case studies from Sri Lanka and Indonesia. *International Journal of Water Resources Development*, 25, 479–489.

Mizushigen Kaihatsu Koudan (1979). *Sameura Dam Kouji Shi* (Records of Construction of Sameura Dam). Tokyo: Kyodo Print.

Nam Theun 2 Project (2004). *Summary Environmental and Social Impact Assessment.* Vientiane: Nam Theun 2 Hydroelectric Power Project.

Nam Theun 2 Project (2005). *Nam Theun 2 Hydroelectric Project: Summary of the Concession Agreement.* Vientiane: Nam Theun 2 Hydroelectric Power Project.

Okawa Mura (1981). *Okawa Mura Shiryou, Vol. 3* (Okawa Village Document). Okawa Village, Japan: Okawa Mura.

Okawa Mura Shi Tsuiroku Hensan Iinkai (History of Okawa Village Revision Committee) (1984). Okawa Village, Japan: Okawa Mura Shi Tsuiroku Hensan Iinkai.

PPLH Unhas (Pusat Penelitian Lingkungan Hidup Universitas Hasanuddin) (1998). *Annual Environmental Monitoring Report.* Makassar: Government of the Republic of Indonesia, Ministry of Public Works, Directorate General of Water Resources Development.

PT. Andel Persada (2006). *Environmental Impact Assessment Report of the Tommo Irrigation Project.* Mamuju: Government of West Sulawesi Province.

Søftestad, T. (1991). Anthropology, development, and human rights: the case of involuntary resettlement. In E. Berg (ed.) *Ethnologie im Wiederstreit. Kontroversen über Macht, Geschäft, Geschlecht in fremden Kulturen.* Munich: Trickster, pp. 365–387.

Takesada, N., Manatunge, J., and Herath, I. L. (2008). Resettler choices and long-term consequences of involuntary resettlement caused by construction of Kotmale Dam in Sri Lanka. *Lakes & Reservoirs: Research & Management*, 13 (3), 245–254.

World Bank (1998). *Recent Experiences with Involuntary Resettlement: Brazil-Itaparika.* Report No. 17544, Operations Evaluation Department. Washington, DC: Author.

World Bank (2004). *Third Mahaweli Ganga Performance Report, Sri Lanka. Project performance evaluation report, Sector and Thematic Evaluation Group.* Washington, DC: World Bank.

Yoshida, H., Shirai, S., Yamazaki, Y., Suda, M., Doi, N., Shimomura, Y., and Fujikura, R. (2010). *Indonesia Bili-Bili Dam itenjumin no kurashi ni kansuru ichikosatsu* (Living conditions of resettlers from submerged area of the Bili-Bili Dam in Indonesia: an analysis with sustainable livelihoods approach). *Ningen Kankyo Ronshu*, 10 (2), 75–90.

Box 3.1 Sri Lanka's Mahaweli Development Program as a framework

Proposals for development were vigorously pursued in Sri Lanka during the 1950s, shortly after the nation's independence. There was very little industry in the country and hence development of agriculture was the main focus. Large-scale water-resources development projects were seen as having the largest potential, and subsequently, Gal Oya and Mahaweli Ganga Development Programs were identified as priority projects.

The Mahaweli Development Program was formulated in the 1960s as a 30-year master plan, and was expected to provide land for the landless, increase food security, and create more employment opportunities in non-farm sectors, together with water-resources development targeting irrigation, hydropower, and flood control. The Program focused on attracting direct official development assistance and multi-lateral development assistance.

The first phase of the Program, between 1969 and 1977, consisted of two projects: construction of the Polgolla Barrage across the Mahaweli River and a tunnel to divert water to what was known as the Dry Zone, and the construction of the Bowatenna Dam and another tunnel to take water to downstream areas.

During the 1970s, annual growth of agricultural production was less than population growth, real per capita gross domestic product (GDP) declined by 1.3 percent annually, and unemployment increased. In 1977, the government decided to accelerate the Program, which envisaged development of 365,000 ha of land for agriculture in 13 systems identified in the Mahaweli Master Plan in the Dry Zone. The total project area was divided into different zones, or systems which were to be developed gradually. During the period from 1978 to 1986, major dams such as Kotmale, Victoria, Randenigala, Rantambe, and Maduru Oya were constructed. Large irrigation settlements in System H (Kalawewa), C (Dehiattakandiya), and B (Welikanda area) were commenced and work was carried out successfully.

The criteria for selecting settlers were very clear from the outset. First priority would be given to those who were displaced from reservoir areas and project-affected areas. Afterwards, those with experience in agriculture as share-croppers or laborers who owned no more than 0.4 ha of farmland from other parts of the country were given land. Limited numbers of project-affected families also had the choice of resettling within the riparian areas by receiving productive land plots or migrating to nearby urban areas by receiving home plots for resettlement.

(continued)

(continued)

A complex system of irrigation canal networks was created in the hope that barren terrain would be transformed into fertile and productive agricultural land. This initiative was intended to solve two pertinent challenges: provide irrigated land to famers in southern regions of the country who either did not have access to land at all or who were subsistence farmers with limited amounts of land; and create employment opportunities, as the country was facing acute unemployment rates. Curtailing the importation of food commodities, primarily rice, was also intended as a direct benefit during a time when the global prices of rice were rising.

First, extensive areas of the jungle were cleared and access roads were built. Impressive networks of canals were then built to supply irrigation water, mostly to revitalize ancient small village tanks, around which new settlements were created. Farm families were settled on 1 ha plots (for agriculture) together with a smaller plot (as a homestead), and each block consisted of about 10 to 15 such farms, supplied by a single farm canal. Several such blocks formed one new settlement village, and roads were then built to connect different localities and facilitate the establishment of a market economy in these newly created settlements. The government created new administrative systems to coordinate and support the new initiatives, and in the late 1970s the future homeland for thousands of families was ready for new settlers to build their new livelihoods, adjust to the new environment and acquire new knowledge and skills that were needed to restore their relocated lives. In addition, the projects provided these areas with health, educational, and other social infrastructure facilities so that the living standards of settlers would be comparable to other areas of the country.

Many appraisal and assessment reports highlight the following observations as some of the factors that needed more attention during planning and implementation of resettlement projects to realize the full benefits of development.

The sizes of the land allocated to command areas and hence the number of resettled farm families was less than what was planned. Delays in the provision of social infrastructure deterred the smooth establishment of new communities. The accelerated implementation resulted in reduced preparatory planning and considerable increase in costs and dependency on foreign funding.

The project objective of boosting production of paddy rice and other crops had limited relevance during implementation stages, as self-sufficiency in rice production no longer applied to the more open trade policies that were prevalent after project implementation and in face of falling prices of rice in the world market. Crop diversification would have solved the problem to a certain extent, especially high-value cash crops such as spices. Such diversification did not materialize, however. Much emphasis was given to strengthening of water-user groups responsible for operation and maintenance of canals, which was very notable from the outset of the projects; however, agricultural support and extension services were weak. Most of the resettlers were satisfied with the irrigation works, farmer organizations, input supply, and access to credit. They were dissatisfied, however, with the process and facilitation of marketing of their outputs, which led to their income being highly unpredictable.

One of the main project objectives was to improve rural livelihoods, but the project could barely realize the goals, as the incremental output of paddy fields was not satisfactory and the income from off-farm non-paddy sources was not dependable. Some resettlers had to transfer away their ownership of land for various reasons, including failed crops, though most of them had not received secure land rights, which led to emerging social disparities in many Mahaweli settlements. Landless households multiplied, as the authorities had failed to secure land for the second generation. Dependants of the original settlers were expected to migrate to other areas or find employment outside the agricultural sector and contribute to more multi-faceted local development. During the past two decades, however, some suitable alternatives have been offered for the younger generation to secure opportunities in small and cottage industries, and they have also been finding employment as skilled laborers in nearby cities. At present, most of the settlers' present income is based on mono-cropping of paddy rice, supplemented by wage earning from non-farm work.

The projects achieved most of their targets for physical works, numbers of beneficiaries, and rice production. As planned, the projects gave priority to involuntary settlers who were displaced from areas flooded by reservoirs built for the Mahaweli scheme. Although their income increased gradually, the stability of income was much lower than expected. The involuntary settlers faced numerous difficulties from the outset, due to harsh environmental conditions and poor social infrastructure compared to their original villages, which were located closer to major cities. Problems related to land ownership and a lack of educational opportunities for their children was also seen as major difficulties. For both voluntary and involuntary settlers, the projects turned out to be unsatisfactory in economic terms, not so much due to poor design or implementation, but, rather, because the terms of trade shifted against rice production.

(continued)

(continued)

Most farmers, especially those in the most recently settled areas, still depend heavily on government initiatives and services. For future resettlement programs, more feasible opportunities should be created for future generations. In the Mahaweli scheme, diversification away from paddy farming to other sustainable options would have offered settlers more prospects. Providing proper legal land titles would also have allowed them to gain more from resettlement compensation.

(Jagath Manatunge)

4 Income diversification

Introduction

We investigated 17 dam-construction projects for this research project. Among the resettlers in these cases, the first generation resettled from the ten dams constructed outside of Japan mostly continued to be farmers after resettlement. It is natural that increases of income enhance the level of satisfaction with life after resettlement. Nevertheless, the natural environment of the resettlement areas is not always similar to submerged areas, so the same level of income might not be necessarily secured by farming the same kinds of crops on the same area of farmland. Crop diversification is a feasible option to avoid impoverishment risk or even to increase and stabilize income. If resettlers can secure non-agricultural income sources besides farming, their incomes will likely be larger and more stable.

In this chapter, we compare the cases of two dams constructed in Vietnam, two villages of resettlers from the NT2 Dam in Lao PDR, and two villages of resettlers from the Koto Panjang Dam in Indonesia, to reveal how income diversification contributed to resettlers' income and satisfaction. Next, we examine resettlers' income structure in Koto Masjid, one of the best-off among the Koto Panjang resettlement villages. Finally, we summarize the most effective approaches resettlers have used for securing secondary incomes.

Yali Falls and Song Hinh projects (Vietnam)

Yali Falls Hydropower Project

The Yali Falls Hydropower Project is located in the central highlands of Vietnam, in the Se San River Basin, which forms part of the Mekong River Basin. The project is located mostly in Kon Tum Province and partially in Gia Rai Province. The development of the Yali Falls Hydropower Project was initiated in the early 1960s as one of the four most promising tributary projects on the Mekong River by the Mekong Committee. It had been the subject of many international studies until the 1990s, when peace returned to the area. During the process of project preparation, many international players were involved, especially Electrowatt (a Swiss consultancy) and

Swedco (a Swedish consultancy), and various socioeconomic and environmental aspects of the project were carefully examined, including resettlement.

The construction of the project coincided with a period of high growth in energy demand, and the project was designed with an installed capacity of 720 MW and annual energy production of 3,600 GWh to be constructed, as part of a series of six hydropower dams on the Se San River. Although the construction of the Yali Falls Project was started in 1993, the detailed design process had been in progress during the late 1980s with the active involvement of international financing institutions such as the World Bank and the Asian Development Bank working through the Interim Mekong Committee, but many international institutions could not work in Vietnam, being constrained by an American economic embargo. This could be a reason that the full implementation of the resettlement program for the affected population had not been carried out in accordance with recommendations by international experts.

According to a previous study (Dao et al. 2004), the reservoir impoundment would displace 1,658 families (8,475 people) that lived in various communes in the provinces of Kon Tum and Gia Lai. During this study, it was found that most of the resettlers in Kon Tum Province then lived in nine villages in two communes (Kroong and Ngoc Bay). The average income per household in the Kroong Commune was estimated to be higher than in the Ngoc Bay Commune. In the Kroong Commune (having four villages: Trung Nghia, Thon Hai, Kroong Ktu, and Kroong Klah), the Trung Nghia Village was selected for this study in view of its stable incomes for many resettlers, from diversified sources, especially from rubber and coffee plantations.

The survey in the Trung Nghia Village was conducted in two stages. First, a survey was carried out only for those families that had resettled directly in the village. Second, a survey was carried out for those households of the second generation in order to identify the possible contributions of the younger generations to improved incomes. In total, we studied 190 resettlers, and their average household characteristics and income are shown in Table 4.1. The results could suggest that younger families tend to do better in terms of income.

Table 4.1 Features related to age of household leaders (Trung Nghia Village)

	Age of household leaders			Second generation	Total
	> 60 yrs	*50–60 yrs*	*< 50 yrs*		
No. of households	58	55	56	21	190
Average household size	4.03	4.65	4.45	4.19	
Average income (VND M/month/capita)	0.99	0.89	1.03	1.75	

Source: Authors.

On the basis of the survey, the data are divided into quintiles of income per capita in order to minimize impacts of accuracy of responses in the survey. The results pointed to an obvious improvement in household income and per capita income. As shown in Table 4.2, the increase compared to income before resettlement was found to be remarkable: from nearly five-fold to over tenfold. The highest increase was with the second and third quintiles, i.e., the middle level. For the poorest quintile, the increase was nearly fivefold.

From the results of the survey, the increase of income could be mainly attributed to the change of sources of income. As shown in Table 4.3, the incomes of resettled households were reportedly diversified from mainly paddy rice and cassava before the project to other new sources after resettlement, including husbandry, non-farm businesses, salaried income, and others. It should also be noted that for second-generation families, the percentage of households earning income from non-farming is higher than for first-generation families.

Table 4.2 Changes in income per capita (Trung Nghia Village)

Quintile	Before resettlement	Without second generation		With second generation		
	Income per capita (VND M/ month)	Income per capita (VND M/ month)	Increase from pre-project	Income per capita (VND M/ month)	Increase from pre-project	Increase from first generation
Fifth	0.512	2.20	479%	2.34	510%	106%
Fourth	0.216	1.11	621%	1.25	702%	113%
Third	0.132	0.74	763%	0.85	876%	115%
Second	0.087	0.51	1,023%	0.55	1,097%	107%
First	0.052	0.27	548%	0.29	579%	106%
Mean	0.063	0.97		1.06		109%

Source: Authors.

Table 4.3 Sources of household income (Trung Nghia Village)

	Reported number of households						
	Farming	Husbandry	Non-farm	Salary	Pension	Others	Total no. households
First generation	141	55	29	25	7	22	169
Second generation	17	5	9	6	–	2	21

Source: Authors.

Upon further examination of the farming practices, the change was significant from mainly paddy and cassava to a bigger share of cash crops and industrial crops as shown in Table 4.4. Among the changes, it was found during the interview that planting rubber brought significantly higher incomes, as one hectare of rubber trees could provide an average monthly income of up to VND 5 to 6 million. This explains the important increase in the number of households planting rubber trees from nil before the project to 111 (over 65 percent of the project total number of households). Information on the farming practices of the second-generation resettlers confirms the important contribution of farming in diversifying sources of income, as mentioned earlier.

Song Hinh Hydropower Project

The Song Hinh Multipurpose Project is located within Phu Yen Province on the southern central coast and situated within the Song Hinh River Basin, forming part of the Song Ba River Basin. The project aimed to divert water into the Song Con River before returning to the Song Ba River. Following approval of the project by the government in 1993, the construction started in November 1995 and the power plant with an installed capacity of 70 MW started electricity generation in April 2000, to generate an average annual energy production of about 370 GWh.

A case study conducted by the Ministry of Industry (MOI) of Vietnam in 2006 provided the first set of findings on results for the Song Hinh resettlement (MOI 2006). The case study pointed out that because the Song Hinh Reservoir is of a medium scale and is located in a mountainous area, the size of the submerged area did not lead to significant negative impacts. The population affected by the reservoir consisted of different ethnic groups, such as Cham, Hroi, Bana, and Ede, of which the latter was the largest, accounting for 46 percent of the total population affected. The area had not benefited from technological advances, lacked food security, and was characterized by poverty and a low quality of life. In total, 473 households were

Table 4.4 Changes in farming practices (Trung Nghia Village)

	Number of households practicing related farming types							
	Paddy rice	Cassava	Sugarcane	Rubber	Fishery	Coffee	Others	Total no. of households
Before resettlement	158	144	8	0	0	0	1	169
First generation	137	13	0	111	15	7	2	169
Second generation	14	2	0	15	2	0	0	21

Source: Authors.

to be resettled, with about 2,000 people in four communes: Duc Binh Dong (163 households), Ea Bia (6 households), Ea Trol (170 households), and Song Hinh (134 households).

On the one hand, the case study pointed out that the resettlement in Song Hinh was considered to be generally successful in terms of better housing and infrastructure provided. In addition, it was reported that apart from compensation for resettlement, with housing, infrastructure, and subsequent mitigation measures provided, the living standard of the resettlers was said to have improved.

On the other hand, independent of the MOI study, Jaakko Pöyry Consulting, a Finnish forestry consultancy, conducted an environmental review of the project and pointed out two negative aspects of the resettlement program:

1   The land compensation provided an average of only 2.37 hectares per household (less than 0.5 hectares per person). In addition, villagers would receive compensation for only about 15 percent of the total area flooded by the reservoir and no compensation whatsoever was proposed for flooded grazing land, forest land, or fallow land.
2   Jaakko Pöyry estimated USD 2.19 million would be required for resettlement and compensation of 299 families, but no explanation was provided why this estimate was less than half that of the World Bank's earlier estimate of USD 5 million in another study in 1993.

According to previous studies and consultations with local authorities, the Song Hinh resettlement started in 1994 and was completed in 2000. Among the ten villages in four communes, the largest resettlement group is located at the Buon Thung Village of the Duc Binh Dong Commune. This village was selected in consultation with the local authorities for the survey, in view of the fact that it is situated on the edge of the reservoir and the relocated settlement is not far away from the location of the original village. Apart from the provision of housing and infrastructure reported by the MOI, support was also provided on cultivation of industrial crops, such as sugarcane and cassava, with a sugar factory and facilities to produce monosodium glutamate for processing cassava. In addition, support was provided for the introduction of green pepper cultivation and aquaculture techniques.

The survey was conducted in two stages, similar to the approach adopted for the Yali Falls Project: first, a survey was conducted for the families directly resettled from the old village; and second, for the second-generation resettlers who now live separately from their parents.

From the two stages of survey, the first findings are related to the ages of the household leaders, as shown in Table 4.5. The household sizes become smaller with younger generations, varying from three to five persons, different from the findings in the Yali Falls Project, with a household size of around four. In addition, the average monthly income is greater than the first group, being 115 percent to 202 percent greater.

Table 4.5 Features related to ages of household leaders (Buon Thung Village)

Song Hinh resettlement	Age of head of household			Second generation	Total
	> 60 yrs	*50–60 yrs*	*< 50 yrs*		
No. of households	31	45	84	22	182
Average household size (persons)	5.00	5.16	4.31	3.23	
Average income (VND M/month/capita)	0.33	0.38	0.50	0.67	
Income increase from first group (%)	–	115	152	202	

Source: Authors.

With respect to the introduction of aquaculture, a series of experiments was conducted with support from the local youth movement to link with tourism development. The model reportedly failed for various reasons, among them the lack of sustained professional support for aquaculture, and the lack of active participation of the local population.

Similar to the analysis of information for the Yali Falls study area, data are divided into quintiles of income per capita in order to minimize impacts of accuracy of responses in the survey. Due to the lack of responses on information of incomes before the resettlement, in view of the similarity of economic situation prior to development as discussed at the beginning of this report, the data of the Yali Falls study were therefore adopted for the income per capita before resettlement. The results presented in Table 4.6 point to an improvement in household income and income per capita, but not at the same extent as for the Yali Falls study area. The highest increase was found

Table 4.6 Changes in income per capita (Buon Thung Village)

Quintile	Before resettlement	Without second generation		With second generation		
	Income per capita (VND M/month) (assumed same as Yali)	Income per capita (VND M/ month)	Increase from pre- project	Income per capita (VND M/ month)	Increase from pre- project	Increase from first generation
Fifth	0.512	1.12	243%	1.25	271%	112%
Fourth	0.216	0.39	216%	0.41	229%	106%
Third	0.132	0.30	309%	0.31	317%	103%
Second	0.087	0.23	469%	0.23	468%	100%
First	0.052	0.14	284%	0.15	294%	103%
Mean	0.063	0.44		0.47		108%

Source: Authors.

with the second and third quintiles, i.e., the middle level. For the poorest quintile, the increase was found to be nearly a tripling of income.

There was no significant change in sources of increased incomes as found with the Yali Falls study area (Table 4.7). The main sources of income were reported to be farming and husbandry, although husbandry could be a significant new source of higher income. In the Song Hinh study area, the second-generation families appeared to be more conservative than those in the Yali Falls study in terms of diversification of sources of income and this fact points to the similar patterns of income between the first- and second-generation families.

In terms of the farming practices, there was a major shift from mainly paddy rice and corn planting to cassava and sugarcane, as shown in Table 4.8. The change reflected the introduction of two new factories for sugar and cassava processing. The change was thus recognized by most farmers as a chance for stable and reliable incomes. However, during one of the surveys, it was reported that the operations of these factories would need improvement to minimize the impacts of middlemen in controlling commodity prices. In the study area, there were efforts to introduce other crops

Table 4.7 Sources of household income (Buon Thung Village)

	Reported number of households						
	Farming	Husbandry	Non-farm	Salary	Pension	Others	Total no. households
First generation	157	19	0	1	0	2	160
Second generation	22	15	0	0	0	0	22

Source: Authors.

Table 4.8 Changes in farming practices (Buon Thung Village)

	Number of households practicing related farming types							
	Paddy	Corn	Cassava	Fruit	Sugarcane	Pepper	Others	Total no. of households
Before resettlement	144	73	9	5	0	0	2	160
First generation	14	2	159	0	18	5	1	160
Second generation	0	1	22	0	4	0	0	22

Source: Authors.

that would allow farmers to deal directly with the market. Peppers were found to be much more profitable and led to a significant increase in monthly incomes. A few examples in the study area pointed to a monthly income up to VND 6 to 8 million. In these efforts, technical know how has been transferred and replicated.

Economic impact of resettlement

During the past three decades, the country enjoyed a period of high economic growth that led to the reduction of poverty in Kon Tum Province (Yali Falls case study) from over 32 percent before 2000 to 22.7 percent in 2012. In Phu Yen Province (Song Hinh case study) the poverty rate went from over 16 percent before 2000 to 15.69 percent (17.7 percent for rural areas) in 2012. On the one hand, a significant income increase for resettlers at both Yali Falls and Song Hinh study areas (as shown in Tables 4.10 and 4.14 later in the chapter) could be mostly attributed to the rapid economic growth in these regions. Nonetheless, the diversified farming practices at Yali Falls contributed to a further increase in income. On the other hand, the unsuccessful introduction of aquaculture, combined with commodity price controls for cassava and sugarcane, hindered the diversification of income in Song Hinh, resulting in a smaller income increase than at Yali Falls. As a result, only the first quartile in Yali Falls is below the poverty line of household monthly income (VND 500,000 per capita), while up to the fourth quintile are remaining below the poverty rate.

NT2 Dam (Lao PDR)

This section focuses on the cases of two resettlement villages, Sophia and Sopphene, among the 16 resettlement villages connected with the NT2 Hydropower Project in the Nakai Plateau in Lao PDR (see Chapter 3). Sophia (336 people) and Sopphene (269 people) were selected because these villages exhibit striking differences in terms of social and economic status eight years after resettlement, although the housing compensation packages and rehabilitation programs were the same. In fact, the average annual income in Sophia (USD 1,854) is almost twice that of Sopphene (USD 1,056).

We sought to understand the role that multiple opportunities for income generation play in resettlement compensation and for resettled villages to sustain livelihoods. This section attempts to highlight underlying reasons for the differences in socioeconomic status of the resettled villages from NT2. On the one hand, according to an interview with the Nam Theun 2 Power Company (NTPC), the resettlement villages of Ban Done, Thalang, Nongbouakham, and Nakai Tai (Neua) are seen as most well-off. On the other hand, the resettlement villages Bon Sopphene, Nongboua, and Sopphene are performing less well in economic and income status among the 16 resettlement villages. We focused on Sophia, which was listed as mid-level in terms of economic performance, and Sopphene, one of the poorest villages, to examine the causes of the gap between the two villages. This section will look into the role played by non-farming income-generation activities.

The old Sophia village was resettled in a new settlement named Sophia village (same name). This village was resettled in 2006. Before the resettlement, it consisted of 56 families, with 336 people. After resettlement, 11 families had chosen to merge with other resettlement villages, 6 families chose to live outside resettlement villages, and 4 other families came from another resettlement village. During the research, the Sophia resettlement village consisted of 43 families and 262 people (129 being female). Five ethnic minorities live in this village: Mand (22 families); Maeu (9 families); Heuheu (6 families); Bor (4 families); and Phong (2 families).

The old Sopphene village was resettled in a new settlement named Sopphene village (same name). This village was resettled in 2006. Before the resettlement, it consisted of 57 families, with 269 people (125 being female). After the resettlement, no families came from other resettlement villages. Three ethnic minorities live in this village: Kleng (26 families); Makong (23 families); and Bor (8 families). During the survey, a sample of 60 households (30 from Sophia and 30 from Sopphene) was selected from 43 and 57 households, respectively, representing 70 percent and 53 percent of the total number of households in the villages investigated.

Household interviews were conducted in Sophia and Sopphene on May 5, 2013, using questionnaire forms and raising the questions one by one to each villager. From each village, 30 households were recommended by the respective village heads, considering the feasibility for the households to take part in such a study. Follow-up interviews with each village head were carried out in 2014 on August 11 and August 13, respectively.

The questionnaire was composed of eight parts: (1) occupation/income; (2) farming activities; (3) property and housing; (4) family; (5) general satisfaction; (6) compensation; (7) culture/traditional relationship between ethnic minorities; and (8) thoughts about children's future.

Occupations of the resettlers after relocation

The majority of resettlers are engaged in farming, fishing, or forestry. All of the households interviewed in Sophia and Sopphene consider themselves to be "self-employed farmers." In Sophia, the highest ranking occupations are as follows (multiple responses accepted): (1) self-employed farmers (100 percent); (2) handicrafts (100 percent); and (3) service (20 percent). In Sopphene, the highest in ranking are as follows: (1) self-employed farmers (100 percent); (2) share-croppers (56 percent); (3) handicrafts (16 percent); and (4) service (10 percent). In Sopphene, 56 percent of the households identify themselves as "share-cropper," down from 100 percent before resettlement. In Sophia, none of the villagers identify themselves as "share-cropper" now, whereas before resettlement, 63 percent of the households considered themselves "share-cropper." This mainly refers to the work-sharing among farmers in the fields of other households in high season. The labor provided is often compensated by barter for food stuffs. The occupations identified do not necessarily mean the activities are compensated in commercial terms. For instance, although 100 percent and 16 percent of households in Sophia and Sopphene, respectively, indicated being engaged in handicrafts, most of them are producing handicrafts primarily for their own consumption.

Increase in income levels and narrowing the disparities

As shown in Table 4.9, the monthly income of both villages more than doubled from LAK 527,000 to LAK 1,236,300 per month in Sophia (2.35 times) and LAK 330,600 to LAK 705,000 per month in Sopphene (2.13 times) respectively. The income in Sophia is almost twice that of Sopphene (1.75 times), mainly due to the lower baseline of Sopphene prior to resettlement. Although the compensation packages were provided with basically the same conditions, Sophia performed slightly better in terms of increasing

Table 4.9 Monthly income by household in Sophia and Sopphene

Family monthly income (LAK)	Before (n = 30)				Present (n = 30)			
	Sophia		Sopphene		Sophia		Sopphene	
	HH	%	HH	%	HH	%	HH	%
150,000–250,000	0	0	9	30	0	0	0	0
260,000–350,000	3	10	9	30	0	0	0	0
360,000–450,000	7	23	6	20	0	0	2	7
460,000–550,000	9	30	4	13	0	0	4	14
560,000–650,000	6	20	2	7	0	0	5	17
660,000–750,000	3	10	0	0	0	0	7	23
760,000–850,000	2	7	0	0	0	0	6	20
860,000–950,000	0	0	0	0	3	10	1	3
960,000–1,050,000	0	0	0	0	5	17	2	7
1,060,000–1,150,000	0	0	0	0	6	20	1	3
1,160,000–1,250,000	0	0	0	0	5	17	1	3
1,260,000–1,350,000	0	0	0	0	4	13	0	0
1,360,000–1,450,000	0	0	0	0	6	20	1	3
1,460,000–1,550,000	0	0	0	0	1	3	0	0
Average (LAK/month)	527,000		330,600		1,236,300		705,000	
Average (LAK/year)	6,324,000		3,967,200		14,835,600		8,460,000	
Average (USD/month)	66		41		155		88	
Average (USD/year)	792		492		1,854		1,056	

Note: HH stands for households in this and subsequent tables.

Source: Authors.

overall income. In fact, the average incomes of both villages are far above the national poverty line of LAK 186,000 per capita per month. When examining the disparities in both villages, the coefficient of variation of income has decreased from 0.2598 to 0.1477 and from 0.3782 to 0.3300, in Sophia and in Sopphene, respectively. In both villages, the gaps between relatively rich and poor have narrowed, and the extent of improvement is far more significant in Sophia than in Sopphene. It seems that Sophia has managed better in terms of both expanding the overall wealth in the village as well as narrowing the gap between rich and poor through the support of the NT2 resettlement project. However, the ratio of lowest to highest range of income in Sopphene before resettlement (LAK 150,000 to 250,000; 560,000 to 650,000) compared to after resettlement (LAK 360,000 to 450,000; 1,360,000 to 1,450,000) has increased from 2.95 to 3.47. This indicates that only a small group of the extreme rich in Sopphene increased income much faster than the others. Actually, the richest household in Sopphene is as wealthy as the richest households in Sophia.

Sources of income

The majority of households in both Sophia and Sopphene earn a substantial amount of income from forestry (100 percent of households), livestock, non-timber forest products, and fishing (Table 4.10). The project supported the marketing of timber from forests managed by a community forest association. However, the company, which used to purchase timber products

Table 4.10 Income sources by household in Sophia and Sopphene

| Income sources | Before (n = 30) | | | | Present (n = 30) | | | |
| | Sophia | | Sopphene | | Sophia | | Sopphene | |
	HH	%	HH	%	HH	%	HH	%
1 Farming	22	73	23	76	9	30	6	20
2 Fishing	2	6	12	40	15	50	17	56
3 Stock farming	13	43	13	43	13	43	17	56
4 Livestock	25	83	24	80	20	66	18	60
5 Non-timber forest products	19	63	20	66	18	60	21	70
6 Hunting	15	50	16	53	0	0	0	0
7 Salary/wage	0	0	0	0	0	0	0	0
8 Handicrafts	30	100	3	10	30	100	10	33
9 Pension	0	0	0	0	0	0	0	0
10 Service	0	0	0	0	4	13	1	3
11 Forestry	0	0	0	0	30	100	30	100
12 Others	0	0	0	0	0	0	0	0

Source: Authors.

from the resettlers, has failed to make payments to the association in recent years, and the income flow has stopped. Fishing is a new and significant source of income for most resettled villages, including Sophia and Sopphene. Due to inundation by the dam, fish yields have increased, and the improved road connections have helped make fishing a good source of marketable products. In theory, the fishery association of each village would purchase the catch from villagers at a fish collection site on the shore of dam lake, and the bulk would be shipped to urban areas such as Thakek. Villagers themselves also visit markets in Oudomsouk to sell their fish.

Table 4.11 indicates the property possessed by households. These assets have financial, social, and cultural significance for the households. In 2013, the inflation rate in Lao PDR was 6.4 percent (World Bank 2014). Property ownership also reflects changes in lifestyle. Villagers now have access to electricity provided by the NT2 Project, and therefore, the majority of households now possess electrical appliances, such as television (color or black and white), telephones, refrigerators, electric fans, satellite receivers, and CD and video players. The villages were provided with facilities for drinking water and attached toilets, which have contributed to the decrease of waterborne diseases among villagers. Regarding vehicles, 50 percent of households in Sophia and 10 percent in Sopphene now possess cars, whereas none of the households had cars before resettlement. With improved road quality and connections, transportation has improved significantly.

Table 4.11 Family property by household in Sophia and Sopphene

| Family property | Before (n = 30) | | | | Present (n = 30) | | | |
| | Sophia | | Sopphene | | Sophia | | Sopphene | |
	HH	%	HH	%	HH	%	HH	%
1 TV (color)	0	0	0	0	26	86	22	73
2 TV (black and white)	0	0	0	0	0	0	0	0
3 Telephone	0	0	0	0	27	90	22	73
4 Radio/cassette	15	50	13	43	9	30	11	36
5 Bicycle	16	53	18	60	13	34	14	46
6 Motorcycle	13	26	9	30	25	83	22	73
7 Car	0	0	0	0	15	50	3	10
8 Refrigerator	0	0	0	0	26	86	25	83
9 Attached toilet	0	0	0	0	30	100	30	100
10 Electric fans	0	0	0	0	21	70	16	53
11 Agricultural tractor	5	16	4	13	4	13	7	23
12 Brass ornaments	10	33	5	16	22	73	18	60
13 Satellite receiver	0	0	0	0	26	86	22	73
14 CD/video player	0	0	0	0	5	16	4	13

Source: Authors.

Project support and general satisfaction

The overall resettlement implementation period designed by NTPC in consultation with the resettlers is still ongoing, and the handover in 2015 to the Nakai District office is approaching, with about 60 staff engaged in the NT2 resettlement. Since 1996, various resettlement support and compensation programs were provided.

When resettlers were asked by the authors during the fieldwork, if they were "satisfied with the place where they live," everyone answered "satisfied." The main reasons they gave were improvements in housing, infrastructure, and access to schooling, etc. When they were asked if they were satisfied with their jobs, the majority of them answered "satisfied." In Sophia, 93 percent of respondents answered "satisfied" and 7 percent answered "don't know." In Sopphene, 100 percent answered "satisfied." Common complaints overall were with regard to the limited size of land areas for farming.

Inter-generational gaps in perception

According to the fieldwork conducted by the authors (Table 4.12), in fact, the majority of the second-generation resettlers desire to be specialists in the future, such as doctors, teachers, police, lawyers, and so on. None of the respondents answered that they would like to be farmers or fishermen in future, although some of their parents hope that they will carry on farming if they have enough land and the means to continue. As the children and grandchildren will have higher levels of education than previous generations, their potential to be engaged in non-farming and higher paying occupations is higher than for the first generation of resettlers. With the urbanization rate in Lao PDR at 4.41 percent (higher than in Thailand) (CIA 2014) and the younger generation not wishing to practice farming, the typical land-for-land approach may not be sufficient to support the rehabilitation of livelihoods of resettlers in the long run.

Table 4.12 Desired future occupations of second-generation resettlers

Number of households				
Ranking	Sophia		Sopphene	
1	Doctor	6	Teacher	11
2	Teacher	5	Doctor	5
3	Police	5	Lawyer	5
4	Lawyer	5	Police	4
5	Artist	4	Soldier	2
6	Businessman	4	Artist	2
7	Journalist	1	Engineer	2
No respondents	Farmer or fisherman			

Source: Authors.

Access to resources

One critical issue is the access to resources. The resettlement project provided each household just 0.66 ha of land for farming, so access to land is limited. Due to the small size of the land provided, to meet the dietary needs of the families, intensive agricultural practices with high soil fertility and irrigation is typically necessary, but this was not the case here. The Sophia villagers were originally from lowland ethnic groups and more used to intensive agricultural practices such as rice growing, and thus coped better with the assistance from the project. However, the villagers in Sopphene were originally from highland ethnic groups practicing slash-and-burn farming. In this sense, the project package did not take into consideration the previous farming methods of Sopphene village, and the transition was not simple. In 2013, one of the project managers of NTPC explained that the goal of the project was to increase the rice yield in resettled villages to 1200 kg/ha. When visited later in 2014, they admitted that the goal was not realistic, and they were looking into other livelihood approaches, such as fishery marketing.

The support and resources provided by the project were basically the same for all villages, including land, housing, infrastructure, livestock, credit schemes, etc. However, Sophia now has better access to high-quality fish resources due to its northern location, close to a conservation area. The baselines of financial, human, and cultural resources form the foundations for development. Some village chiefs pointed out that the villages that are better-off today, in fact, already started with better financial assets compared to others, and this also made a difference in post-resettlement performance. These points can be observed from Table 4.11. One major concern is that a "land-for-land" scheme itself is not enough, and there is a need to pursue and promote multiple income-generating activities apart from farming.

Adaptive capacity and market orientation

A second issue is how to promote multiple income-generating activities in addition to farming. Households in both Sophia and Sopphene earn a substantial amount of income from forestry (100 percent of households), livestock, non-timber forest products, and fishing. One of the major differences is that 100 percent of households in Sophia identify handicrafts as a source of income, but only 33 percent in Sopphene. The traditional weaving practiced by female villagers in Sophia has been marketed with the support of the project. They are innovative and forthcoming in exploring new business opportunities such as handicrafts, which were made only for their own consumption prior to resettlement. As Sophia is making substantial income from handicraft products, villagers in Sopphene are also slowly learning from Sophia's success and trying to identify products such as their bamboo woven baskets to be developed as marketable handicraft products. Despite gaps among villages, there is mutual learning among them through exchanges of information, prompting villages to apply elements of success from others. As with the size and quality of land provided, villagers can no longer maintain subsistence-based agriculture and livelihoods. There is a tendency for resettlers who are adaptive, by harvesting and developing marketable products, to be better-off than others in these two villages.

Human capital as a basis for development

Looking into the human factor of the two villages under study, the baseline of human capital and the consequent learning curves show some differences. The compositions of the ethnic groups in each village were described above. Sophia has five ethnic minorities, and all the families are identified as lowlanders. Sopphene has three ethnic minorities, and most families are identified as highlanders. Cultural traits might characterize the highlanders. Village members indicated that "they will be happy as long as they have enough for today," implying a weaker sense of investing in the future. They were also isolated and had less exposure to learning from external experts before resettlement. The NTPC staff members observe less motivation and initiative from Sopphene compared with Sophia villagers.

Human capital corresponds to any stock of knowledge or characteristics a worker has (either innate or acquired) that contributes to his or her "productivity." These include school quality, training, and attitudes toward work, etc. Education, knowledge, skills, competencies, attitude, health, and welfare have an impact (Acemoglu and Autor 2014). Although there is limited data on the baseline of attainment of education in these villages, the literacy level of the older generation is reportedly slightly higher in Sophia than in Sopphene. However, overall, the primary and secondary enrollment rates have improved significantly for both villages. Therefore, the second-generation resettlers now have access to better education due to project support and have greater opportunity to further their education and seek higher-paying jobs apart from farming than the first generation. In just a few years, there could be a strong workforce to support the livelihoods of the first generation.

Conclusions

A number of criticisms of the NT2 dam were about the short-term perspective. When the time horizon is looked at in the longer term, there are gradual processes of adjustments. For instance, a number of problems pointed out by Matsumoto and Harashina (2012) were solved by initiatives of resettlers and the project. One major concern is the practice of so-called "land-for-land" schemes, providing farmers alternative lands after resettlement. In NT2, it appears that land provided as compensation was not sufficient for resettlers. Villagers in Sophia are taking advantage of other income-generation activities (e.g., handicraft production), and producing substantial income. Sophia appears to have stronger human capital (education) as a basis to gain more from the project support (e.g., handicraft marketing). To narrow the gap, more support may be needed, responding to the needs of the respective villages. The needs may change and evolve through exposure to more information and interactions with other villagers. Inter-generational issues are also of significant concern. More opportunities should be given to children. The villagers do not wish their children to continue farming, and neither do the children themselves. Resettlement compensation only in the form of land may be fine for the first generation, but not for the second. This is something to be considered in future policy-making. Although project support ends in 2014, the productive capacities of resettlement villages are not yet restored, and thus, a continuation of assistance from the project or the district is necessary.

Koto Panjang Dam (Indonesia)

In Chapter 3, we described impoverishment in two resettled villages in West Sumatra Province, Tanjung Pauh and Tanjung Balik. Meanwhile, there are better-off villages in Riau Province after resettlement. In this section, we compare one best-performing village from resettlement (Koto Masjid, Riau Province) with one worst-performing village (Tanjung Balik, West Sumatra). Then, we introduce another study investigating factors that made Koto Masjid economically active.

Comparative study on Koto Masjid and Tanjung Balik

Methodology

There are two objectives of the study: (1) to clarify the present living conditions of resettled residents compared to their living conditions prior to resettlement; and (2) to associate the present living conditions with some cause so that inferences can be made concerning how to improve resettlement planning in the future. The village selection is mainly based on income improvement after resettlement (see Chapter 3). A random-sampling method was used to select 50 resettled households from each village to participate in a survey. All interviews were conducted in December 2010. Because the focus of this study was to evaluate the performance of involuntary residents, only resettled families were interviewed. If by chance a non-resettled household was contacted, the interviewers skipped to the next family until 50 resettled families in each village had been interviewed.

Findings

The Koto Panjang resettlement villages have been in operation for about two decades. The relocations took place between 1991 and 2000. The first relocation was to Pulau Gadang in 1991 and to Tanjung Balik in 1992. The final relocation to Pulau Gadang was in 1996 and to Tanjung Balik in 2000.

The majority of the relocations took place between 1992 and 1993. The two villages in West Sumatra to be resettled were Tanjung Balik and Tanjung Pauh. The Riau authority has transformed many villages by implementing a specific strategy for rural development. Koto Masjid is a division of the village of Pulau Gadang.

Construction of the Koto Panjang Dam was part of an Indonesian development program to generate energy from its expansive water resources. The consequences of the dam construction included the relocation of nearly 5,000 households. The success of the resettlement was due to the prospect of economic growth in their new villages. The government promised to provide improved community facilities and infrastructure as well as financial compensation and homes for every resettled household. This was enough to convince the natives of Koto Panjang to relocate. This study has validated the fulfilment of those promises; all of the relocated individuals now enjoy improved community facilities and infrastructure.

Pulau Gadang, the village where the residents of Koto Masjid once lived, was completely submerged at the beginning of the Koto Panjang Reservoir inundation; Tanjung Balik, however, is not completely submerged. Settlers now living in Koto Masjid are no longer able to visit their old village as a result of this submersion. Settlers native to Tanjung Balik are still able to visit their old villages and still work with the rubber trees that they had to leave behind. The different opportunities offered to the two villages might explain their current performance. Residents of Koto Masjid have no choice with respect to their past; they must develop their destiny in a new home. Their background might no longer hinder their current and future lives. In contrast, residents in Tanjung Balik might still have their current and future life tied to the old village in some way. Thus, residents of Tanjung Balik may not fully develop their resources in the new village because they still have opportunities in their old village.

The Koto Panjang resettlement was carried out following the perspectives of Cernea (1999). However, the decision to relocate was not necessarily forced. Table 4.13 classifies the degree of freedom perceived by residents when the resettlement option was offered. If they decided to relocate on their own, they are classified as voluntary resettlers. If they agreed reluctantly or had to relocate, they are classified as involuntary resettlers. Involuntary

Table 4.13 Agreement to accept resettlement (percent)

Did you ever agree?	Village		Combined
	Koto Masjid	Tanjung Balik	
Yes	58	16	37
Yes, but reluctantly	26	64	45
Didn't agree	16	20	18

Source: Authors.

resettlers accounted for 63 percent of the individuals in Koto Panjang; voluntary resettlers accounted for 37 percent.

The original character of the residents differentiates Koto Masjid from Tanjung Pauh. Voluntary resettlers in Koto Masjid account for 58 percent of individuals; involuntary resettlers account for 42 percent. In contrast, voluntary resettlers account for 16 percent of individuals in Tanjung Balik; involuntary resettlers account for 84 percent. According to Cernea (1999), the risk of impoverishment is always present in involuntary resettlement. Therefore, the potential risk of impoverishment is more serious in Tanjung Balik than in Koto Masjid.

There are several factors that motivate families to accept a resettlement program. In the case of Koto Panjang's residents, the most important motivating factors were land ownership, a new home, cash compensation, and a good location. The degree of influence of each motivating factor is the same for Koto Masjid and Tanjung Balik. Other non-physical motivating factors, including business, crafts for children, and education for children were not as important. The degree of influence of each of these non-physical motivating factors was also the same for Koto Masjid and Tanjung Balik.

The authors find that the resettlement process was successful. The process entailed a planning stage and many details leading up project implementation. The government successfully motivated the local natives of Koto Panjang to relocate by providing better possibilities for living.

Koto Panjang's resettlement program used various types of compensation to persuade local communities to relocate. Compensation included cash, new housing, land, a plantation, education and health facilities, and social infrastructure. Many of the residents disagreed with the compensation, but the majority did not complain. It seems that the main complaint was that the amount of cash compensation did not satisfy many of the residents; only 14 percent of the respondents openly accepted everything about the resettlement program. They report not feeling satisfied with the compensation in general but they did nothing to get their complaints heard. Only 19 percent of respondents sent their complaints to the village head and only 12 percent of respondents sent their complaints to government officials. Less than 1 percent took their case to court. The difficulty of raising official complaints is one of the social problems perceived by residents in the two villages.

Resettlers responded differently to the compensation. In Koto Masjid, 28 percent of resettlers stated that they were satisfied with it. In contrast, no resettlers in Tanjung Balik were satisfied with the compensation i.e., no resettlers expressed agreement with it.

Occupation

Most respondents surveyed indicated that they worked on their own farm both prior to and after resettlement. The importance of a self-employed farm is related to the presence of a rubber plantation. The number of respondents working on their own farm has increased, from 81 percent prior to resettlement to 88 percent after resettlement. The proportion of respondents in Koto Masjid working on their own farm increased from 80 percent prior to resettlement to 90 percent after resettlement; the proportion in Tanjung Balik increased from 82 percent prior to resettlement to 86 percent after resettlement. Other occupations that resettled residents are engaged in include share-cropping, working in the public sector, and small trading. However, these occupations are becoming insignificant lately because residents are working more on their own farms.

Sources of income

It is evident from Table 4.14 that the rubber plantation is becoming quite important in the economy of Koto Panjang. Rubber as the primary source of income has contributed to an increase in the standard of living and is restoring the main economic activity of residents. The role of the rubber plantation as the primary source of income was indicated by 79 percent of respondents prior to resettlement and 80 percent after resettlement. In terms of Koto Masjid, this marks an increase from 78 percent prior to resettlement to 90 percent after resettlement, while in Tanjung Balik there was a decrease from 80 percent prior to resettlement to 70 percent after resettlement. The Koto Panjang economy traditionally depends on the rubber plantation. In line with the development of a dominant economic structure, the Koto Panjang resettlement program planned a rubber plantation of 2 ha for every resettled

Table 4.14 Resettlers' primary sources of income (percent of respondents)

Source of income	Village		Combined
	Koto Masjid	Tanjung Balik	
Before resettlement			
Rubber plantation	78	80	79
Public salary	6	0	3
Trading	6	12	9
Wage labor	8	8	8
Others	2	0	1
After resettlement			
Rubber plantation	90	70	80
Fishing	2	2	2
Public salary	6	0	3
Trading	2	18	10
Wage labor	0	8	4
Others	0	2	1

Source: Authors.

family. The project expected to transfer an already productive 2 ha of rubber plantation to every resident at the start of resettlement in new villages. Residents expected to own productive rubber trees in their new village.

However, the above expectation appears to have been empty. The planned rubber plantations did not materialize due to poor project monitoring. This disappointed many residents and led to a public outcry to the authorities responsible for the resettlement project. The Indonesian government eventually responded to the disappointment caused by this implementation failure (see Chapter 3). A new rubber plantation was financed by a national budget in both Tanjung Pauh and Tanjung Balik. The present rubber plantation in Tanjung Balik became productive much later than in Koto Masjid. The rubber plantation is expected to increase in importance as the primary source of income for residents in Tanjung Balik, as it is now for residents in Koto Masjid.

The return of the rubber plantation as the primary source of income is a stabilizing feature in the economy of Koto Panjang. Rubber is a major export commodity of Indonesia, and increased trade liberalization will open a larger market for rubber and bring higher prices. This is good news for local farmers in the resettlement villages.

Other income-earning activities include trading, wage labor, public employment, and fishing. Among these activities, trading is the most important as a primary source of income. Prior to resettlement, trading was an important source of income for 9 percent of respondents. Since resettlement, this has increased to 10 percent. The role of trading in Koto Masjid fell from 6 percent of respondents prior to resettlement to 2 percent after resettlement, reflecting the increasingly dominant role of the rubber plantation. The role of trading in Tanjung Balik is evident from its rise as the primary source of income for 12 percent of respondents prior to resettlement to 18 percent after resettlement. The strategic position of Tanjung Balik (on the national highway connecting Padang, the capital of West Sumatra Province, and Pekan Baru, the capital of Riau Province) has been instrumental in expanding trade activities.

Secondary sources of income

The availability of a secondary source of income remains important, even if it is decreasing, for residents in the Koto Panjang resettlement economy. Table 4.15 shows the decreasing availability of a secondary income source from 62 percent (total having a secondary source, in "Combined" column) of respondents prior to resettlement to 49 percent of respondents after resettlement. This trend is true for Tanjung Balik, but not for Koto Masjid. The availability of a secondary income source in Tanjung Balik fell from 74 percent of respondents prior to resettlement to 34 percent after relocation. The availability of a secondary income source in Koto Masjid rose from 50 percent of respondents prior to resettlement to 72 percent after resettlement.

Table 4.15 Resettlers' secondary sources of income (percent of respondents)

Source of income	Village		Combined
	Koto Masjid	Tanjung Balik	
Before resettlement			
Not available	50	26	38
Rubber	10	12	11
Fishing	4	2	3
Public salary	4	0	2
Trading	8	20	14
Wage labor	16	4	10
Gambier	0	36	18
Others	8	0	4
After resettlement			
Not available	36	66	51
Rubber	0	4	2
Fishing	36	4	20
Public salary	4	0	2
Trading	12	6	9
Wage labor	0	2	1
Gambier	0	2	1
Food agriculture	6	14	10
Palm oil	2	0	1
Cattle	4	2	3

Source: Authors.

The composition of secondary income sources suggests that household incomes for residents in Koto Masjid are more stable than for residents in Tanjung Pauh, because more residents in Koto Masjid have a secondary income than in Tanjung Balik. Furthermore, the sources of income are more diversified in Koto Masjid than in Tanjung Balik.

A diversification of economic activity is taking place in both Koto Masjid and Tanjung Balik; this is obvious from the variety of secondary income sources. These income sources are mostly based on new products since resettlement and are almost completely different from the secondary income sources prior to resettlement. This observation holds true for both Koto Masjid and Tanjung Balik. The role of fishing as a secondary income source in Koto Masjid increased substantially, from 4 percent of respondents prior to resettlement to 36 percent after resettlement. Koto Masjid is now famous all over Riau Province for its catfish and smoked fish. Koto Masjid also supplies fish products to other regions. The rise of the fishing industry in Koto Masjid is due to its local water resources and the increasing demand for fish in Pekan Baru and Padang. In addition to fishing, palm oil is providing a secondary income for residents of Koto Masjid. The dynamic local resettlement economy in Koto Masjid is obvious from the presence of new buildings and new cars that belong to the residents. Local economic performance in Koto Masjid is also attracting many agencies, especially the well-financed provincial government of Riau and private companies attempting to benefit from the growing local market.

In contrast, Tanjung Balik belongs to the less advantaged provincial government of West Sumatra. The local land is suitable only for food agriculture and gambier. Gambier is the dried sap extracted from leaves and small twigs of the gambier tree and has multiple applications, including tanning and dyeing substances, pharmaceuticals, leatherworking, and in the textile industry. Gambier is an export commodity of Indonesia and thus very important for the region (Lima Puluh Koto Regency) as a whole, especially the villages of Tanjung Balik and Tanjung Pauh. Foreign investment in gambier processing has made it even more important for the Tanjung Balik economy. Gambier is easy to sell because the processing company accepts gambier leaves directly from local farmers. There are three gambier-processing factories in the region, owned by Indian investors. The most accessible factory for residents producing gambier is located in Batu Bersurat, Kampar District, on the banks of the Koto Panjang Reservoir.

Family income levels

The level of income earned dictates a resident's capacity to purchase things required to support daily living. The question was whether residents are making progress in terms of their individual purchasing power. Unfortunately, there is no statistical information on household income in these resettlement villages. They were asked about their household income based on the current price level, and asked to judge their income prior to resettlement based on the price index at the time of the survey (2010). Basically, the aim was to determine their household income prior to and after resettlement in order to compare the two.

The distribution of household income levels is presented in Table 4.16. Household income is classified from the lowest to highest, from IDR 500,000 or less per month to IDR 4,500,000 or more per month. The percentage of respondents with a household monthly income below IDR 500,000 per month fell from 43 percent prior to resettlement to 3 percent after resettlement. The percentage with a household monthly income between IDR 500,000 and IDR 1,000,000 fell from 39 percent to 14 percent. The percentage with a monthly income between IDR 1,000,000 and IRD 1,500,000 rose from 10 percent to 13 percent. The percentage with a monthly income between IDR 1,500,000 and IDR 3,000,000 rose from 8 percent to 44 percent. No respondents received a monthly income above IDR 3,000,000 prior to resettlement; however, after resettlement this accounts for 26 percent of households.

A majority of respondents have experienced an increase in real income. Prior to resettlement, 82 percent of respondents earned less than IDR 1,000,000 per month; after resettlement, 83 percent of respondents earn more than that. This indicates that an improved income distribution within the resettlement economy

Table 4.16 Resettlers' household income levels (percent of respondents)

Household monthly income (000s IDR)	Village		Combined
	Koto Masjid	Tanjung Balik	
Before resettlement			
< 500	42	44	43
500–1,000	40	38	39
1,000–1,500	8	12	10
1,500–3,000	10	6	8
After resettlement			
< 500	4	2	3
500–1,000	4	24	14
1,000–1,500	4	22	13
1,500–3,000	46	42	44
3,000–4,500	16	4	10
> 4,500	26	6	16

Source: Authors.

has accompanied the increase in real household income. In other words, the process of economic growth is functioning and diminishing the impoverishment risk.

Koto Masjid and Tanjung Balik are both experiencing an improved income distribution and an increase in real household income. The proportion of respondents in Koto Masjid earning less than IDR 1,000,000 per month accounted for 82 percent prior to resettlement. After resettlement, 92 percent earn more than this. This is also taking place in Tanjung Balik. Prior to resettlement, 82 percent of respondents earned less than IDR 1,000,000 per month. After resettlement, 72 percent earn more than this. Household income levels have increased since resettlement.

Living conditions

Improved living conditions are important in reducing the risk of impoverishment associated with involuntary resettlement, and have improved for more than 70 percent of respondents; specifically, 88 percent of respondents in Koto Masjid and 52 percent of respondents in Tanjung Balik note an improvement in their living conditions. An increased proportion of respondents enjoying better living conditions indicates strong economic performance.

Conclusion

The process of economic development following the resettlement program of the Koto Panjang Dam project was stimulated with monetary compensation, increased productive capacity, and an increased level and distribution of income. Better-off villages received a higher level of compensation and used the compensation to purchase more productive assets. More productive capacity (i.e., rubber plantations) was found in better-off villages. Also, household income is greater and is substantially increasing in better-off villages. Household income in the worst-off villages is slightly increasing in one village but is significantly decreasing in another village. An increase in the level of household income is followed by better income distribution and a lower level of poverty, whereas a decrease is followed by worse income distribution and an increased level of poverty. An improved standard of living is associated with better productive capacity, higher income, better income distribution, and a lower level of poverty. The risk of impoverishment is lower in the better-performing villages, reflecting the better condition of their productive capacity. The presence of productive capacity is necessary to guarantee the success of an involuntary resettlement program whose goal is to improve the standard of living for displaced peoples.

Intensive study on Koto Masjid

Following the above mentioned study, we conducted an intensive study on Koto Masjid, comparing the performance of the village through time by using data collected from household surveys in 2004 and 2013. The study included a personal interview of 50 households in 2004, and 100 households in 2013. The selection of households in the two surveys has consistently followed a systematic random sampling.

The study also used the secondary data on population and employment available from the Indonesian Statistics Agency, and from the Koto Masjid Village Office (Koto Masjid 2013). Since household income collected from the survey is in nominal value that follows prevailing prices at the survey date, the calculation of household income in real value results from the use of a GDP Price Deflator published by the Indonesian Statistics Agency at 2000 constant prices (BPS 2014a). The availability of real household income figures in 2004 and 2013 makes the measurement of economic growth possible for Koto Masjid.

Population growth

The population of Koto Masjid increased from 1,235 in 2000 to 1,855 in 2010 and 2,068 in 2013. The increase in population has affected the number of households. The number of households increased from 259 in 2000 to 416 in 2010, and 535 in 2013. As a result of faster growth rate of population than the growth rate of households, the average size of families decreased from five to four members.

The population growth rate in Koto Masjid has showed a declining trend (4.1 percent for 2000–2010 and 3.6 percent for 2010–2013), but it is well above the population growth rate of Indonesia, which was less than 2 percent in the same period. A higher rate of population growth results from the higher birth rate and inflows of migration. The high population growth rate in Koto Masjid implies increasing prosperity. Growing prosperity in the local village economy enables families to have more children and attracts inflows of migration. Since the risk of impoverishment is inherent in involuntary resettlement, the high population growth rate in Koto Masjid may indicate that resettled families have passed Scudder's third stage of the resettlement process (Scudder 2005).

Structure of employment

Agriculture is the main source of livelihood in Koto Masjid, as shown in Table 4.17. The whole agricultural sector accounts for 73 percent of respondents. Rubber plantations are dominant in the agricultural sector, and account for 60 percent. Rubber plantations are the legacy of past generations. Rubber trees grew naturally in the forest, and local people only tapped the trees, without planting them. Rubber tapping has been a traditional economic activity for generations. Changing from an economic tradition can be risky. Therefore, with the approval of local people the resettlement program included the development of a rubber plantation for families displaced due to the development of the Koto Panjang Dam.

The restoration of rubber plantations as the main source of livelihood encouraged the development of catfish farms. The entire fishery sector in Koto Masjid originates from home-based catfish farms, which contribute 11 percent of the respondents' employment. The role of catfish farms will continue to increase in the future. In response to local initiatives in the promotion of the catfish fishery, the central government has selected Koto Masjid village as a center for the development of catfish farms in the province of Riau. The government has proved its commitment by building facilities and infrastructure to enhance the competitiveness of the catfish industry initially developed by local communities. The national government has constructed a business center for smoked catfish production, and it is processing raw materials of catfish from local producers. A vertically integrated relationship has naturally developed in the catfish business. The catfish producers benefit from the presence of the catfish smoking center, and the center benefits from the presence of local producers that guarantee the supply of catfish raw materials. The center provides the equipment needed to process smoked catfish, ranging from cleaning the fish to the smoking kitchen, and warehouse to store the smoked fish products, and ready them for marketing. Although the marketing of smoked catfish targets markets in the city, consumers can buy smoked catfish directly from the center. The presence of catfish farming encourages the development of a processing industry and trading.

Table 4.17 Number of respondents in Koto Masjid, by main and additional employment in 2013

Main employment	Additional employment									Total
	None	Rubber plantation	Fishery	Industry	Trading	State employee	Private employee	Others	Total	
1 Rice farming	0	0	0	0	0	0	1	1	2	2
2 Rubber plantation	32	2	11	2	2	1	3	7	28	60
3 Fishery	1	8	0	0	1	0	0	1	10	11
4 Industry	0	1	2	0	0	0	0	0	3	3
5 Trading	3	2	0	0	0	0	0	0	2	5
6 Public employee	2	1	3	0	0	0	0	3	7	9
7 Private employee	1	0	0	0	0	0	0	0	0	1
8 Others	2	2	3	1	1	0	0	0	7	9
Total	41	16	19	3	4	1	4	12	59	100

Source: Authors.

The entire proportion of households with additional employment is 59 percent of the total. The most important contribution of additional jobs comes from the catfish farming and rubber plantations. The role of the catfish fishery is more important as an additional employment than as the main employment. However, both rubber plantations and the catfish fishery also play a significant role in the creation of main employment as well as additional employment.

Economic structure

The increasing presence of non-agricultural activities is evident from the household survey conducted in 2013. Table 4.18 shows some economic indicators in the local economy of Koto Masjid. The indicators include household income, employment structure, and economic structure. The agricultural sector remains significant in employment and in contributing income, but the role of non-agricultural sectors are starting to show promise. Within the Koto Masjid agricultural sector, the role of rubber plantations remains crucial as a source of livelihood improvement. Rubber plantations account for 60 percent of total employment in Koto Masjid. The next significant economic activity goes to fisheries. The fisheries sector provides employment for 11 percent of working households. The catfish farm business sector dominates the fishery. The Koto Masjid village government has a vision to promote the development of catfish ponds in every house. "No home without catfish pond" is the official motto for Koto Masjid village development.

Table 4.18 Household income, sectoral share of employment and income in Koto Masjid in 2013

Main employment	Average household income (IDR/year)	Share (%)	
		Employment	Income
1 Agriculture	97,931,342	73.00	58.12
1.1 Rubber plantation	79,565,400	60.00	38.81
1.2 Fishery	189,940,364	11.00	16.99
2 Non-agriculture	190,811,556	27.00	41.88
2.1 Industry	523,400,000	3.00	12.76
2.2 Trading	163,920,000	5.00	6.66
2.3 Public employee	108,594,667	9.00	7.95
2.4 Private employee	48,000,000	1.00	0.39
2.5 Others	192,973,333	9.00	14.12
Total	123,009,000	100.00	100.00

Source: Authors.

The local economy of Koto Masjid is making a transition toward having a greater role for non-agricultural sectors. The dominant role of the agricultural sector is empowering the growth of non-agricultural sectors. Accumulated savings from rubber plantations supply the need for money for starting businesses outside agriculture. The industrial sector started to create employment opportunities in the rural economy. The growth of the catfish fishery sector leads to local industrial growth. It is producing fish-feed industry. Local fish-feed producers can sell at a price much lower than the big factory-made fish feed. Local fish-feed quality is not lower than factory-made fish feed.

The regional market welcomes the entrance of catfish producers from Koto Masjid. The business environment makes catfish farming competitive and profitable commercially. The growth of catfish fisheries encourages the growth of fish-feed producers, a smoked catfish industry, and catfish fingerling businesses. The economic linkages of catfish fishery are significantly wide starting upstream with the wealth of water resources to downstream with the presence of the catfish business development center.

The ongoing transformation of the local economy in Koto Masjid demonstrates the economic power of integration where water resources act as a driver. The economic transformation starts from the wealth of ground water resources. It is the resettled families that make themselves a subject of development, rather than an object of impoverishment. The resettlement project planned only rubber plantations for the source of livelihood. The rubber plantations provided the resettlers with the seeds of capital to develop the local economy on their own. Lately, the growth of autonomous local efforts is successfully attracting external support from the government. The gain is the inflow of capital to Koto Masjid.

The dynamics of the local economy in Koto Masjid reflect the standard of living achieved by resettled families. In 2013, the average household income of resettled households reached IDR 123 million. It is indicating that average resettled family in Koto Masjid is more prosperous than the average agricultural household in Indonesia. The Indonesian Census of Agriculture in 2013 reports that the average income of agricultural households has only reached IDR 12.4 million a year (BPS 2014b). The average agricultural household in Koto Masjid is evidently much richer than the average agricultural household in Indonesia. This is logical and reasonable because the average resettled household in Koto Masjid owns more land than the average agricultural household in Indonesia. The average farm household in Koto Masjid owns more than 2 ha of land, whereas the average farm household in Indonesia owns less than 1 ha (BPS 2013). The difference in the size of land ownership is resulting in the difference in average income, and differences in the standard of prosperity. Land ownership for the agricultural sector is still capable of making prosperity for farmers. This is clearly evident in Koto Masjid, an involuntary resettlement caused by the dam development in Koto Panjang. The evidence also supports that land for compensation is still relevant for displaced families due to development projects in the agricultural dominant economy.

Off-farm household income proved to be much higher than the income of farm households. In the agricultural sector, fisheries households generate an average income of about IDR 190 million a year. In non-agricultural economic activities, industry sectors provide the greatest household income and trade activities. This picture shows the economic linkages between sectors, particularly fisheries, industry, and trade. The roots of the economic linkages are a wealth of water resources that have been able to transform society from the risk of impoverishment due to forced resettlement into an optimistic and creative community to develop new economic activities. Sustainability of water resources is urgent for the development of Koto Masjid local economy. The development requires government policy to sustain water resources as a strategic resource that is capable of integrating the transformation of the local economy.

Economic transformation tends to occur more rapidly in the structure of income than in the structure of employment. In the structure of employment, the agricultural sector provides employment for 73 percent of the working households, while in the structure of income, the agricultural sector contributes only 58 percent of the entire income received by households. At the same time, the non-agricultural sector, which provides employment for 27 percent of working households, contributes 41 percent of the total income received by households.

Economic growth

The rate of economic growth is measured by using household income data from the study conducted in Koto Masjid in 2004 and 2013. Household income follows prices prevailing at the time of the survey. Since the measurement of the growth rate requires the data for real income, the national GDP Deflator at year 2000 constant prices is used to get real income of households. Table 4.19 shows the monthly real income of households by economic sector for 2004 and 2013 based on year 2000 constant prices. The calculated real income provides data to measure economic growth for over the last nine years in the rural economy. Stable economic growth can help households avoid the risk of impoverishment.

The average annual rate of economic growth in Koto Masjid was 9.3 percent between 2004 and 2013. The economic growth results from the contributions of the agricultural and non-agricultural sectors. The rates of economic growth show differences across sectors. Non-agricultural sectors as a whole achieved an average annual economic growth rate of 14 percent, while the agricultural sector was at 9.5 percent. However, the performance of the agriculture remains significant in the rural economy of Koto Masjid, and the contribution of agriculture increased from 54.5 percent of community total real income in 2004 to 58.1 percent of the village total real income

Table 4.19 Income, structural change, and economic growth in Koto Masjid

Household economic activity	Monthly household real income (IDR)		Income share (%)		Economic growth: 2004–2013 (%/year)
	2004	2013	2004	2013	
1 Agriculture	1,057,563	2,488,849	54.49	58.12	9.51
1.1 Rubber plantation	1,018,294	2,022,093	46.47	38.81	7.62
1.2 Fishery	–	4,827,187	–	16.99	–
2 Non agriculture	1,358,613	4,849,327	45.51	41.88	14.00
2.1 Industry	–	13,301,804	–	12.76	–
2.2 Trading	1,956,159	4,640,564	25.92	20.78	9.60
2.3 Public employee	2,547,015	2,759,849	18.75	7.95	0.89
2.4 Private employee	577,227	1,219,883	0.85	0.39	8.31
2.5 Others	–	–	–	14.12	–
Total	1,358,649	3,126,178	100.00	100.00	9.26

Source: Authors.

in 2013. The strength of agriculture in the economy of Koto Masjid reflects the rise of catfish-farming activities. In 2004, there was no household interviewed yet having catfish farming as a source of income. But in 2013, the contribution of catfish accounted for almost 17 percent of the population's total real income. At the same time, the rate of economic growth in the rubber plantation activity is only 7.6 percent, much lower than the economic growth rate for the rural economy. The lower rate of economic growth made the role of rubber plantations fall from 46.5 percent down to 38.8 percent of village total real income. The rate of economic growth in the catfish fishery has implicitly exceeded the rate of economic growth in agriculture to support the increasing role of agriculture in the rural economy of Koto Masjid. It also indicates that structural change has taken place within agricultural sectors.

The rise of the catfish fishery coincides with the rise of industry and trade. The growth rate of trading accounts for 9.6 percent, but the contribution fell from 25.9 percent of the village total real income in 2004 to 20.8 percent of the community total real income in 2013. Like catfish farming, in 2004 no household yet had industrial activities as a source of income. In 2013, the role of the industrial sector suddenly appeared to account for 12.8 percent of the village total real income. Implicitly, the appearance of the industrial sector indicates that the sector performed well to support the higher growth rate in the non-agricultural sector. The economic linkage of catfish fishery to the industrial sector and trading spurs economic growth and structural change in the resettlement local economy. The presence of catfish fishery has contributed to the economic transformation of Koto Masjid.

Inequality impact of economic growth

High economic growth requires a reasonably equitable distribution of income to maintain economic stability. In 2004–2013, the Koto Masjid economy revealed a high level of economic performance. The Koto Masjid economic growth is much higher than the national level of economic growth. Indonesia was achieving 7 percent in economic growth after the financial disaster in the late 1990s, but the average rate of economic growth in Indonesia in the period 2004–2013 fell to 5.8 percent (BPS 2014c). Does a high rate of economic growth support more equitable distribution of income? Using household real income in 2004 and 2013, Table 4.20 shows average household real income and economic growth by ten equal proportions of households (deciles) from first to tenth. The first decile is the poorest, and the tenth the richest.

All elements in the society enjoyed economic growth during the years 2004 to 2013. However, the rate of economic growth differed by decile. For the first and second deciles, the economic growth rates were 8.8 percent and

Table 4.20 Real income, growth, inequality, and living conditions in Koto Masjid

Deciles of households	Average real income (IDR)		Growth (%)
	2004	2013	2004–2013
1	202,030	444,037	8.75
2	339,121	834,095	10.00
3	570,012	1,047,269	6.76
4	663,811	1,297,711	7.45
5	721,534	1,501,981	8.15
6	808,118	1,746,018	8.56
7	995,717	2,095,484	8.27
8	1,529,652	2,809,695	6.76
9	2,164,602	3,811,676	6.29
10	5,591,889	15,673,815	11.45
Average real income	1,358,649	3,126,178	9.26
Gini coefficient	0.51	0.56	
Improved living condition? (%)			
Yes	66	94	
No	34	6	

Source: Authors.

10 percent, respectively. The rate of economic growth was 6.8 percent in the third decile, and higher from the fourth to the seventh decile, ranging from 7.5 percent to over 8 percent. Although the economic growth rate declined from the eighth to the ninth decile, it is still above 6 percent, and the tenth decile shows the highest growth rate at 11.5 percent.

Although the rate of economic growth by class in society does not vary sharply, the gap in real income per household shows a wide gap between the richest and poorest decile. In 2004, the poorest decile received real income per household of up to IDR 0.2 million. At the same time, the richest decile received real income per household of up to IDR 5.6 million. In 2013, the poorest decile received real income per household of up to IDR 0.4 million, and the richest decile received real income per household IDR 15.7 million. The ratio of real income per household between the richest to the poorest households has increased from 28 times in 2004 to 35 times in 2013. The gap between the most affluent segment and the poorest segment in the society is enormous, and it is widening as the rural economy grows.

Along with the high rate of economic growth, the Gini coefficient increased from 0.51 in 2004 to 0.56 in 2013. The Gini coefficient for Indonesia was lower than for Koto Masjid in the same period. Indonesia's Gini coefficient is on an increasing trend from around 0.32 in 2004 to 0.41 in 2013 (BPS 2014d). Both Koto Masjid and Indonesia are facing the challenge of rising income inequality.

Changes in living conditions and prosperity

Information on the state of livelihood came from a questionnaire used for a household survey. The survey has documented information with a first survey in 2004, a second survey in 2010, and a third survey in 2013. The survey asked the households to confirm "Yes" or "No" for comparing the living condition in the resettlement villages with the living condition in the original village. If their living condition is better now than before, the answer will be "Yes." If their living condition is worse now than before, the answer will be "No." Table 4.20 shows the results for living condition from the surveys in 2004 and 2013. The proportion of households indicating improved living condition in the resettlement village increased from 66 percent in 2004 to 94 percent in 2013. This indicates that livelihood improvement has taken place. Positive and high economic growth in the dominant sectors during the period 2004 and 2013 also provides an indication to confirm improved livelihoods.

Conclusion

The agricultural sector remains significant in employment and in contributing income, but the role of non-agricultural sectors is starting to show much promise. Within the agricultural sector, the role of rubber plantations remains crucial as a source of livelihood improvement. The next significant economic activity is fisheries. The local economy is transforming toward a greater role for non-agricultural sectors. The dominant role of the agricultural sector is empowering the growth of non-agricultural sectors. Accumulated savings from rubber plantations are supplying the need for capital to start businesses outside agriculture. The industrial sector has started to create employment opportunities in the rural economy. The growth of the catfish fishery sector is leading to local industrial growth. The business environment is making catfish farming more competitive and profitable commercially. The economic linkages of a catfish fishery are significantly broad, starting upstream with the wealth of water resources to downstream with the presence of a catfish business development center. The ongoing transformation of the local economy demonstrates the economic power of integration where water resources act as a driver. The economic transformation starts from the wealth of ground water resources.

Discussion

In all of the cases in Vietnam, Lao PDR, and Indonesia, it was observed that the more diversified the income sources, the more incomes increase. It is also likely that the diversification of income sources contributes to income stability. In order to diversify income sources, implementation of some measures may be necessary. They include guidance from implementing agencies to resettlers and/or governmental policies. In Vietnam, the national government policy promoting crop diversification has been implemented at a resettlement site. It likely contributed to an increase of resettlers' incomes.

In NT2, the implementing agency actively provided support to secure secondary income sources, but the effectiveness of measures differed among villages. Different ethnicity and lifestyles before resettlement caused differences in effectiveness and income increases.

In Koto Panjang, a comparison between the best-off and the worst-off villages reveals the effectiveness of the introduction of a fishery as a secondary income source. In Koto Masjid, the best-off village, employment has been successively generated by fish farming, fish-feed production, and fish smoking. The village was able to obtain more income than Indonesia's national average. This successful economic development was based on villagers' entrepreneurship and capital accumulation, accomplished by rubber production on farmland provided as resettlement compensation. Support from the government by establishing a fish-processing facility also facilitated the business.

Support from implementing agencies and/or governments is crucial for resettlers to secure and enhance secondary incomes. Whether the economy of a resettlement village develops successfully depends not only on entrepreneurship of the resettlers but also the socioeconomic situation. It is very difficult to forecast the long-term effectiveness of support at the planning stage. The implementing agencies should monitor economic activities on a long-term basis and provide additional support if necessary.

Another important element to realize diversified incomes is the establishment of reliable road networks and means of communication both within the resettlement villages and outside of the project command area.

In both the Koto Panjang and NT2 cases, access from the project command area including resettlement villages to major cities was significantly improved thanks to the development of access roads for the sake of dam construction. The success of aquaculture in the Koto Masjid village was apparently supported by improved access to the nearby major city of Pekanbaru, where there is a demand for fresh fish. Income from resettlers fishing on the reservoir in the NT2 case is also supported by the much-improved access to the nearby large city of Thakek, thanks to the construction of the roads within the project command area and to Thakek. In both cases, the roads originally developed for dam construction are still fairly well maintained, and proved pivotal to provide the resettlers with access to the outside, particularly to nearby major cities.

Better access to nearby large cities also leads to secondary development in and around the project command area. A number of small shops and restaurants were developed in the downstream area of the Koto Panjang Dam for customers who came mainly from Pekanbaru at weekends. This development relies much upon good access to the oil-rich city of Pekanbaru, about one hour away by road.

To a lesser extent, secondary development for tourists has also been observed in and around the resettlement villages created by the NT2 dam construction. In the town of Nakai, which is located in the center of the NT2 resettlement villages, some secondary development has been observed for Lao people and travelers from abroad. Lodging facilities were also developed in the resettlement villages. Improved access to Thakek and to Thailand through a bridge over the Mekong River has created such secondary development.

Improved communication among resettlement villages, both by improvement of road traffic and information and communication technologies (the proliferation of mobile phones in particular) also contributed significantly to the diversification of income for resettlers. The NT2 case clearly shows that increased exchanges of information among resettlement villages and resettlers have led to the creation of secondary sources of income for resettlers. The establishment of reliable road networks and means of communication with outside of the community were instrumental for success of aquaculture and fishing in the resettlement villages of the Koto Panjang case.

In the Vietnamese cases also, improved and reliable communication with the outside should be an important element for crop diversification, for its success relies much on the information about the markets of the products. Improvements in road networks, which took place along with the construction of dams, also guarantee diversification of crops in the context of faster delivery of fresh agricultural products to market.

The fact that the second generation of resettlers prefers non-farming occupations may also stem from improvements in road traffic and communication. It seems fair to assume that the resettlers now live in a much more "open" environment with the outside of their own villages after relocation, as compared with the villages being fairly "closed" before resettlement. The second-generation villagers are exposed to more information, much of it first-hand information from neighbors or themselves, regarding life outside their native villages. They no longer automatically expect to take over the occupations of their parents, because they are aware of other vocational opportunities for their future.

Whether these changes are really good or bad for a community is an open question. Some may think that traditional cultures, inherited over generations, are threatened by extinction, and that this is a real threat to communities. Others may think that such changes are inevitable, and can take place in association with changes in society caused by resettlement, and that they, at the same time, offer new opportunities for the second-generation villagers. The value judgment should be left to the resettlers. It should be stressed, however, that decisions by the second-generation resettlers should also be respected, and that measures should be taken to assist the older generation to the extent possible, for they may be the last resort for the region's culture, which might disappear otherwise.

References

Acemoglu, D. and Autor, D. (2014). *Chapter 1: The Basic Theory of Human Capital.* Lectures in Labor Economics. Retrieved from http://econ.lse.ac.uk/staff/spischke/ec533/Acemoglu%20Autor%20chapter%201.pdf.

BPS (Badan Pusat Statistik) (2013). *Agricultural Census 2013.* Retrieved from http://st2013.bps.go.id/.

BPS (2014a). *Gross Domestic Products from Badan Pusat Statistik.* Retrieved from www.bps.go.id/.

BPS (2014b). *Census of Agriculture and Operating Income Household Survey of Agriculture 2013.* Jakarta: BPS.

BPS (2014c). *Growth Rate of Gross Domestic Product.* Retrieved from www.bps.go.id/.

BPS (2014d). *Selected Consumption Indicators, Indonesia 1999, 2002–2013.* Retrieved from www.bps.go.id/.

CIA (Central Intelligence Agency) (2014). *2014 World Factbook Laos and Thailand.* Retrieved from https://www.cia.gov/library/publications/the-world-factbook/geos/la.html and https://www.cia.gov/library/publications/the-world-factbook/geos/th.html.

Cernea, M. M. (ed.) (1999). *The Economics of Involuntary Resettlement: Questions and Challenges.* Washington, DC: World Bank.

Dao, T. H., Dao, T. V. N., and Tran, C. T. (2004). *Study into Resettlement at the Yali Falls Dam, Kon Tum Province.* Hanoi: Institute of Ecology and Biological Resources and International Rivers Network.

Koto Masjid (2013). *Registered Population in Koto Masjid*. Koto Masjid: Village Government.

Matsumoto, I. and Harashina, S. (2012). Effects of livelihood restoration plans on dam project: peoples' livelihoods changes by the Nam Theun 2 Hydropower Project in Lao PDR. *Japan Society for Impact Assessment Journal*, 10 (2), 65–76.

MOI (2006). *Case Study 07-02: Resettlement—Song Hinh Multipurpose Project, Vietnam*. Hanoi: MOI . Retrieved from www.ieahydro.org/reports/Annex_VIII_CaseStudy0702_SongHinh_Vietnam.pdf.

Scudder, T. (2005). *The Future of Large Dams: Dealing with Social, Environmental, Institutional and Political Costs*. London: Earthscan.

World Bank (2014). *Lao PDR*. Retrieved from www.worldbank.org/en/country/lao.

Box 4.1 Turkey's Southeastern Anatolia Project (GAP): dam development to address ethnic issues

The Southeastern Anatolia Project (in Turkish, Güneydoğu Anadolu Projesi, or GAP) of Turkey is a multi-sector and integrated regional development project for the Southeastern Anatolia region, which is on the border with Syria. It was first launched in 1977, and aims at improvement of living standards and income levels of people in the GAP region, so as to eliminate regional development disparities and contribute to such national goals as social stability and economic growth. The GAP region has an area of 75,358 km^2, which corresponds to 9.7 percent of the total territory of Turkey. According to the results of the 2000 General Population Census, the population of the region was 6,604,205, which corresponded to 9.7 percent of the total population of the country.

The project command area of the GAP includes 41.5 percent of the total watersheds of the Euphrates and Tigris Rivers. The project includes construction of 22 dams and 19 hydroelectric power plants on the Euphrates and Tigris and their tributaries for a total installed capacity of 7,500 MW, which represents 22 percent of total electric energy potential in Turkey (Altinbilek 1997; Akca et al. 2013). It aims to irrigate 1.7 million hectares of the area to grow cash crops and promote agri-industries such as food processing for export. The planned irrigation area corresponds to 20 percent of total irrigable land in Turkey. The dams will increase the irrigated land in Turkey by more than 40 percent. The GAP was assumed to generate 3.8 million jobs, mostly in agriculture and raise per capita income in the region by 209 percent (Ronayne 2005). Upon completion of the project, 28 percent of the total water potential of Turkey will be managed through the facilities on the Euphrates and the Tigris (TACC-Southeast n.d.) (Table 4.21).

Table 4.21 Major dams planned within the framework of GAP

Dam	Capacity (million m³)	Power generation (MW)	Year of completion
Euphrates River			
Ataturk	48,700	2,400	1992
Birecik	1,220	672	2000
Karkamis	157	180	1999
Tigris River			
Kralkizi	1,075	94	1998
Dicle	595	110	1999
Kayser		90	Planned
Silvan	6,800	160	Under construction
Batman	1,175	198	2003
Garzan	165	90	Under construction
Ilisu	10,410	1,200	Under construction
Cizre	360	240	Planned

Source: FAO (2009).

Turkey has limited energy resources and is heavily dependent on imported oil and gas. The government estimates that there are potential domestic sources for 246,000 GWh per year of electric power generation, of which 125,000 GWh is by hydropower. Turkey has about 1 percent of the total world hydroelectric potential. The government once hoped to expand hydropower generation capacity to 35,000 MW by the year 2010, while about a half of this target (18,234 MW) was materialized by 2010 (Kaleagasi 2013). Construction of more than 300 additional hydroelectric power plants is projected to make use of the remaining potential hydroelectric sites, about 69,000 GWh per year (USDOE 2003). Hydropower generation is therefore the major component of the GAP.

The GAP was initially planned as an energy and irrigation project to utilize the potential of the rich water and land resources in the region. It was thereafter seen by the Turkish government as a key element to help resolve conflict with Kurdish people in the country (Jongerden 2010). The majority of them in Turkey reside in Southeast Anatolia. The government view is that armed conflict with Kurdish rebels is triggered by poor socioeconomic development of the region, and takes the position that economic development and an increase in investment in Southeast Anatolia may help to integrate the alienated Kurdish population and stop separatist PKK (Kurdistan Workers' Party) activities by

(continued)

(continued)

eliminating the economic motivation of the rebel movement (Carkoglu and Eder 2001). The GAP was cited by the Turkish government as a fulfillment of the Copenhagen criteria for Turkey's entrance into the EU (Hatem and Dohrmann 2013). However, the GAP has been criticized for flooding a significant portion of the historical Armenian, Assyrian, and Kurdish homelands (Oktem 2004). Inundation of the ancient city of Hasankeyf, a cultural significance to the Kurdish people (Berkun 2010), was even criticized as an act of ethnic discrimination or even genocide (Shoup 2006).

The GAP has been expanded since its original design, and now includes such activities as the building of schools, roads, health care centers, housing, cultural projects, women's projects, and the development of tourism (Ronayne 2005). A recently conducted comparative analysis has shown that southeastern Anatolia is still the most disadvantaged region of Turkey in terms of per capita income, per household minimum food expenditures, and per capita cost of basic needs. The Turkish government recently announced a new road map for the social and economic development of the project command area, by investing in the future of the GAP. Political and economic instability in Turkey in the 1980s diverted attention from the GAP. The late 1990s are considered a lost decade for the Turkish economy. The new announcement of a massive investment plan for the GAP has raised expectations and hopes. The new GAP Action Plan is expected to help the region reach its long-awaited normalization while providing the Turkish economy with a strong foundation based on agriculture and energy sectors (Ozhan 2008).

(Mikiyasu Nakayama)

References

Akca, E., Fujikura, R., and Sabbab, Ç. (2013). Atatürk Dam resettlement process: increased disparity resulting from insufficient financial compensation. *International Journal of Water Resources Development*, 29 (1), 101–108.

Altinbilek, H. D. (1997). Water and land resources development in southeastern Turkey. *International Journal of Water Resources Development*, 13 (3), 311–332.

Berkun, M. (2010). Environmental evaluation of Turkey's transboundary rivers' hydropower systems. *Canadian Journal of Civil Engineering*, 37 (5), 684–694.

Carkoglu, A. and Eder, M. (2001). Domestic concerns and the water conflict over the Euphrates-Tigris river basin. *Middle Eastern Studies*, 37 (1), 41–71.

FAO (2009). Euphrates–Tigris Basin. Aquastat. Rome: FAO. Retrieved from www.fao.org/nr/water/aquastat/basins/euphrates-tigris/index.stm.

Hatem, R. and Dohrmann, M. (2013). Turkey's Fix for the "Kurdish Problem." *Middle East Quarterly*, 4, 49–58.

Jongerden, J. (2010). Dams and politics in Turkey: utilizing water, developing conflict. *Middle East Policy*, 17 (1), 137–143.

Kaleagasi, B. (2013). Renewables in Turkey. Presented at European Parliament, Brussels, January 22, 2013. Retrieved from http://brusselsenergyclub.org/get_file/id/b.kaleagasi-turkeyrenewables-short-22january2013.ppsx.

Oktem, K. (2004). Incorporating the time and space of the ethnic "other": nationalism and space in Southeast Turkey in the nineteenth and twentieth centuries. *Nations and Nationalism*, 10 (4), 559–578.

Ozhan, T. (2008). *New Action Plan for Southeastern Turkey*. SETA Policy Briefs Book, 18. Washington, DC: SETA Foundation.

Ronayne, M. (2005). The Cultural and Environmental Impact of Large Dams in Southeast Turkey. Retrieved from www.nuigalway.ie/archaeology/documents/ronayne_turkey_dams.pdf.

Shoup, D. (2006). Can archaeology build a dam? Sites and politics in Turkey's Southeast Anatolia Project. *Journal of Mediterranean Archaeology*, 19 (2), 231.

TACC-Southeast (Turkish American Chamber of Commerce of the Southeast United States) (n.d.). Southeastern Anatolia Project (GAP). Retrieved from www.taccsoutheast.com/pictures/Southeastern_Anatolia_Project.pdf.

USDOE (2003). *An Energy Overview of the Republic of Turkey*, Washington, DC: Office of Fossil Energy, USDOE. Retrieved from www.geni.org/globalenergy/library/national_energy_grid/turkey/EnergyOverviewofTurkey.shtml.

5   Addressing emotional aspects of dam resettlement

Introduction

It is well known that people develop emotional attachments to the places in which they live. Memories of the landscape, nature, family, relatives, and friends seem unforgettable. In particular, one's original home has special meaning where livelihoods depend on natural resources, such as for farmers, fishermen, or foresters. Dam construction submerges a place that is irreplaceable, and the original inhabitants suffer a sense of loss. Even when they are economically better-off after resettling, they often have a sense of loss and feelings of victimization caused by the dam construction.

The emotional issue has not always been adequately taken into consideration in resettlement programs, which have often been designed exclusively from an economic perspective. The emotional issues should not be overlooked, however. The failure to consider these issues may cause not only dissatisfaction among the resettlers, but also may hinder the actual sustainability of the resettlement program itself. In the cases presented in previous chapters, such as the Kotmale (Chapter 3), Saguling (Chapter 2), and Wonorejo (Chapter 2) dam projects, many resettlers chose to resettle near their original homes rather than moving to remote areas.

This chapter presents three cases of dam construction projects: Bili-Bili (Indonesia), Atatürk (Turkey), and Miyagase (Japan). In the case of the Bili-Bili Dam, many who moved to resettlement areas hundreds of kilometers away returned to near the reservoir within a few years after the initial resettlement. Some of them returned because of the hardships in the resettlement areas, but others decided to return despite having relatively better living conditions compared with their original homes. The reason for the latter returning was emotional; they simply wanted to live near their original homes. Resettlement of the Atatürk Dam was conducted without substantial participation of the resettled people. On the one hand, they could not obtain equivalent farmland because land speculation inflated land costs, and this increased the disparities between the rich and the poor among the resettlers. They lamented the loss of opportunities to meet relatives and friends, as well as the local prestige they lost because of the resettlement.

On the other hand, the poor farmers who ventured to move about 1,300 km away to a resettlement area are significantly better-off and positively appreciate the resettlement. Activities related to the Miyagase Dam, such as compensation negotiations and the establishment of resettlement sites, lasted 20 years after the announcement of the dam project due to the full involvement of resettlers, and their compensation was generous. Many preferential treatments were provided to the resettlers in addition to the regular monetary compensation. However, they still had the feeling that they were forced to make sacrifices for the national policy.

Bili-Bili Dam (Indonesia)

As shown in Chapter 3, resettlers joined the TP (see Box 2.2) and moved to Mamuju or Luwu, places distant from their original home, and they faced difficulties. Many of them returned to places near their original home in the district of Gowa. This section analyzes the reasons they returned and discusses how the resettlement should be evaluated.

The number of resettlers and their destination of relocation are summarized in Table 5.1. More than half of the resettled families (1,079 families, 51.7 percent of total) chose to resettle in the vicinity of the reservoir in Gowa District. These particular resettlers were able to purchase land in the vicinity, because the amount of compensation they received was enough to buy other land close to their original homes. They enjoyed improved living conditions with respect to housing and electricity, and were better-off than the original residents, as shown in Chapter 3. Another 415 families (19.9 percent) relocated to urban areas, and most of them quit farming and changed occupation, as shown in Chapter 6.

There were 592 resettled families (28.4 percent) that chose to join the TP. They had owned small plots of land or no land at all, and did not obtain

Table 5.1 Destinations and number of resettled households

Alternative	Destination	Number of registered households	Relocation period
Purchased new land and/or	Gowa District (reservoir vicinity)	1,079	1989–1995
built a new house	Urban areas	415	1989–1995
	Total	1,494	
Joined the TP	Luwu District	200	1990–1991
	Mamuju District	392	1991–1995
	Total	592	
Grand total		2,086	

Source: PPLH Unhas (1998).

sufficient compensation money to buy land in the vicinity. Transmigration areas in Mamuju and Luwu were several hundred kilometers away from their original homes; these residents had almost no option but to join the TP and receive land for living and cultivation.

The Ministry of Transmigration agreed to allocate a certain amount of land for resettlers in the TP areas in the districts of Mamuju and Luwu. The conditions offered to them were identical to those offered to the transmigrants from Java and Bali, but a number of resettlers returned to the Gowa District. Out of the 141 families that resettled to Luwu, only 41 (29.1 percent) remained until 2004. In of the case of Mamuju, 71 of 115 families (61.7 percent) were residing there up to 2007.

Survey results

A survey was conducted targeting resettlers returning from transmigration areas to the reservoir vicinity (hereafter, "returnees") to examine the role of the TP as an additional option in the relocation scheme. A total of 101 returnees were interviewed at six villages in the Manuju and Parangloe subdistricts of Gowa District between December 2010 and January 2011. Returnees in the villages were identified by asking respondents to supply, for the purposes of this study, the names of other returnees they knew. Through this approach, 48 returnees from Luwu and 53 from Mamuju were located. Although no data were available concerning the destinations of resettlers who left the transmigration areas, the number of returnees from Luwu interviewed amounted to nearly half of the resettlers who had left.

The survey covered the following issues:

1   compensation received;
2   reasons for returning to the dam vicinity;
3   living conditions before, during, and after transmigration.

The surveyed returnees were either heads of resettled households or members who had experienced life in a transmigration area. Their ages ranged from 24 to 80, with 58 percent between the ages of 40 and 70. Seven returnees were more than 70 years old. In terms of gender, 67 were male and 34 were female. Most of the returnees had a limited educational background, having completed only primary school or less. The average length of stay in the transmigration area was 3.8 and 2.7 years for returnees from Mamuju and from Luwu, respectively. The combined average was 3.23 years, with 53 percent returning to Gowa District between 1992 and 1995.

The survey results indicated that 79 percent of returnees considered their cash compensation to be in accordance with the promised amount; the remaining 21 percent considered it less than the promised amount. In terms of overall satisfaction, 70 percent considered the amount to be satisfactory, while the remaining 30 percent considered it unsatisfactory. Regarding the various ways they spent their compensation money, 85 percent of returnees purchased homes and/or land where they presently live. Other uses included food, education, and motorcycles. About 10 percent used the money for a pilgrimage to Mecca and 9 percent spent it on wedding ceremonies.

As shown in Chapter 3, resettlers in both Luwu and Mamuju suffered hardships during the first years after resettlement. Most of the returnees from Luwu (40 of 48) stated that their main reason for leaving the transmigration area was land-related problems (Table 5.2). Twenty-four returnees experienced frequent flooding, resulting in decreased crop yield. Sixteen returnees gave land disputes as their main reason. Returnees from Mamuju also suffered from floods, but only a small percentage experienced land disputes. The next most important reason for returning was their desire to reunite with wives and children who had accompanied their husbands to the transmigration areas but immediately returned to their original homelands (14 returnees from Mamuju, 4 from Luwu). Eight returnees from Mamuju stated that health and age-related problems were the main reason for returning. Twenty-one returnees from Mamuju were older than 60, while only seven from Luwu were in that age range.

Most returnees from Mamuju had owned more land in the transmigration area than they did in their original homelands and present location (Table 5.3). In contrast, most returnees from Luwu had owned less land in the transmigration areas. The number of those who were landless in their present location increased among returnees from both Mamuju and Luwu.

A common pattern of changes in occupation was observed among returnees. In their original homes, out of the 101 returnees, 87 were self-employed farmers and 11 were tenant farmers. The number of self-employed farmers increased to 98 in the transmigration areas and then decreased to 64 after returning. Data concerning land ownership, irrigation water access, and occupation imply that self-employed farming was not attractive enough to prevent

Table 5.2 Reasons for returning to the reservoir vicinity (surveyed December 2010 to January 2011)

Reasons	Mamuju	Luwu	Total
Low productivity/income/floods	15	24	39
Land disputes	3	16	19
Insurmountable hardship	6	0	6
Desire to reunite with families remaining in the original vicinity	14	4	18
Livelihood unfavorable/safety reasons	3	4	7
Desire for better access to children's education	4	0	4
Health and advanced age-related problems	8	0	8
Total	53	48	101

Source: Authors.

Table 5.3 Land ownership of returnees (surveyed December 2010 to January 2011)

Land ownership (ha)	Mamuju			Luwu		
	Original vicinity	TP area	After returning	Original vicinity	TP area	After returning
< 2	10	5	4	6	2	4
2	2	32	0	0	0	0
1– < 2	19	7	6	9	2	3
0.1– < 1	17	9	33	32	44	32
0	5	0	10	1	0	9
Total	53	53	53	48	48	48

Source: Authors.

other returnees from leaving the transmigration areas. Most returnees from Luwu worked as public officers, including teachers and policemen. About 40 percent of resettlers in Mamuju continued living in the transmigration area, having improved their income and living standards by growing cash crops. It was possible that some returnees preferred being tenant farmers rather than self-employed farmers; growing new crops is more risky than cultivating rice under the conventional share-cropping system. Although agriculture is still the dominant livelihood in the reservoir vicinity, 22.9 percent of returnees worked in non-agricultural sectors as traders, construction laborers, and in other jobs. This suggests that job opportunities in non-agricultural sectors have increased throughout the past two decades.

In terms of homes for the returnees, 24 percent were in very good condition, 68 percent in moderate condition, and 8 percent in poor condition. On the one hand, when comparing the current conditions to previous ones (both before and during the transmigration period), the current housing conditions are generally far superior. On the other hand, a minority of returnees (14 percent) claimed that their pre-transmigration (original) housing condition had been superior to their current homes. A smaller percentage (3 percent) claimed that the housing conditions during their transmigration period had been superior to their current homes.

A similar tendency was observed regarding utilities and properties/assets. Their current conditions were much better compared to the previous ones; about 52 percent of returnees owned a refrigerator, often indicative of upper-middle-class households. This included 31 of the 53 returnees from Mamuju and 21 of the 48 returnees from Luwu. Owning a motorcycle, as well as the physical qualities of one's home, were often indicators of upper-middle class or upper-class households (which included 56 percent of returnees). Even controlling for the contribution of Indonesia's national economic development during the past decades, it is possible to conclude that the present living conditions of returnees were better than before.

Water access, however, had greatly decreased compared to previous conditions. All returnees had access to wells in the transmigration area, but only 30 percent had such access at the time of the survey; 32 percent now had access to water from the PDAM (Perusahaan Daerah Air Minum), the regional state water company. A similar tendency was observed when it came to irrigation water access: 46 percent of returnees claimed that access to irrigation during transmigration was better than their current access. Only 9 percent claimed to have better irrigation access presently. This water problem was partly attributed to their coming back to the region; at the planning stage of the regional development project, including construction of the Bili-Bili Dam and its irrigation system, there was no consideration of such a large number of resettlers returning to the vicinity.

For the sake of corroboration and cross-checks, the survey asked questions regarding various levels of satisfaction. The responses were as follows: 80 percent were satisfied with their current occupation, 92 percent were satisfied with their current living conditions, and 96 percent had no plan to relocate. Many of the returnees lived in fairly good conditions, 24 percent in very good conditions, and 8 percent in poor conditions. These results were consistent with the level of satisfaction expressed concerning their jobs and living conditions.

Reasons for returning

Among the reasons given for returning to the reservoir vicinity, the most common ones were land disputes and poverty in the TP areas. Their housing conditions and utilities had improved, and many ranked as middle-class citizens in their present villages. Prior to this survey, it was assumed that the returnees were poor, that this was why they had joined the TP, and that they had failed to establish a livelihood in their transmigration area. A deeper analysis of the returnees' behavior (i.e., their purchasing of homes) revealed what may have been the real reasons for returning. Regarding their present homes, 85 percent of returnees purchased them between 1990 and 1999. The remainder purchased their houses within two years of returning to the reservoir vicinity (Table 5.4). About 85 percent of returnees purchased their present homes using their own funds, while 13 percent inherited their homes from their parents or received them from the local government (2 percent moved in with other relatives). This is consistent with their responses concerning spending the compensation money: 64 percent had purchased land and homes, and 21 percent had purchased land only. Some returnees admitted that they had raised additional funds to build their present homes by illegally selling their old homes and land in the TP areas. Only two returnees did not own their own homes, and they were currently living with relatives.

Based on an analysis of the data, a pattern of behavior can be identified for returnees from the transmigration areas:

Table 5.4 Number of relocated project-affected families that purchased their present home, in four time periods

Periods	Mamuju			Luwu		
	Relocated to TP	Returned to reservoir vicinity	Purchased/ built present home	Relocated to TP	Returned to reservoir vicinity area	Purchased/ built present home
1990–1991			5	48	4	8
1991–1995	53	32	18		36	20
1995–2000		14	21		6	11
2001–2006		7	7		2	9
Total	53	53	51*	48	48	48

Note: * Two of the returnees moved in with relatives and do not own their own homes.

Source: Authors.

1   They were all relocated to transmigration areas but returned to the dam's vicinity because they found the conditions in the transmigration area too difficult and/or they wanted to live closer to their families.
2   Those who had received enough cash compensation purchased land and/or homes in their original vicinity.
3   Those who had successfully saved money in the transmigration areas bought land and homes using their compensation cash.
4   Those who had not been able to make enough money in the transmigration areas returned to the dam vicinity and are currently living with relatives.
5   Much of the land in the transmigration areas was illegally sold to others or given to the recipients' children.

The resettlement scheme for the Bili-Bili Dam development provided cash compensation for all resettlers and offered the additional option to join the TP. The cash compensation provided many options for the resettlers, and about 70 percent purchased land close to their original vicinity or relocated to urban areas. Given that the resettlers were not an isolated ethnic minority in the region and also that Indonesia has experienced rapid economic growth throughout the past two decades, even the resettlers who relocated to the reservoir vicinity may still engage in non-agricultural sectors of work.

Cash compensation, however, does not necessarily provide opportunities for the poor. The landless and the small-scale landowners received less compensation, and they may not have been able to restore their quality of life. The TP provided such disadvantaged groups with opportunities to gain new land and to increase their incomes.

While all participants in the TP experienced hardships throughout their first years, 40 percent of resettlers in Mamuju successfully established their livelihoods. Most returnees also obtained land and homes in the reservoir vicinity by taking advantage of the TP, and are satisfied with their present conditions. The operation of the TP at Mamuju and Luwu was not satisfactory, however, and those who participated in the TP suffered various hardships. Those who were able to overcome the difficulties and manage to capitalize on opportunities, however, improved their lives to levels that exceeded their pre-transmigrant lives (Yoshida et al. 2010).

The survey found that 8 percent of returnees lived in poor conditions. This is consistent with findings from other interviews in the reservoir vicinity and the transmigration area of Luwu. These individuals had been landless in their original location and had a limited capacity to make use of new opportunities because of the low level of human, social, and physical capital accessible to them. When faced with land disputes in Luwu, many resettlers returned to the reservoir vicinity; however, those who wanted to return but did not have the money to do so had to remain in Luwu. These individuals are the most vulnerable against major events such as relocations and disputes; they tend to be at the bottom of the social ladder and have no voice in development programs. Careful attention should be paid to this group, and necessary assistance, including vocational training and support for children's education, should be combined with cash compensation and other safety-net programs.

The resettlement scheme was fairly successful, except for a small number of poor families who benefitted from neither the cash compensation nor the TP. The return of some families from the TP areas does not necessarily mean that the resettlement scheme was a failure; the return was mostly due to the poor operation of the TP, which caused many problems for the residents and resulted in some of them illegally selling their land. The diversity of the lives of the resettlers mirrored the demographic change currently taking place in Indonesia; Sulawesi has been experiencing rapid economic growth, and the rapid urbanization of the city of Makassar has increased job opportunities in non-agricultural sectors. More people are encouraging their children to obtain higher education and to work in these urban sectors. In this context, cash compensation was relevant to most of the resettlers. The TP gave the disadvantaged group additional support both directly and indirectly, although there was much room for improvement in its implementation.

Atatürk Dam (Turkey)

The Atatürk Dam is located in the southeastern region of Turkey and was built to meet the country's increasing energy and water demands within the Southeastern Anatolia Project (hereafter, the GAP, the Turkish acronym, see Box 4.1). Resettlement as a result of the construction of the Atatürk Dam began in 1989 and ran until 1991. The land and homes of approximately

55,300 people were fully or partially inundated as a result of this construction; 1 town and 11 villages were fully inundated and 3 towns and 79 villages partly inundated. Out of the 55,300 people, 19,264 from 3,251 households resettled to New Samsat, Hatay, Söke, and Aydın, where new resettlement areas were developed as a part of the governmental project. Others moved to Adıyaman and Kahta, which are close to the inundated settlement sites (Akyürek 2005). Others took their compensation and purposefully moved to Adana and Mersin, some hundreds of kilometers away from their homes (Kadirbeyoğlu 2009). These cities are relatively developed, with a high potential for employment, and have been accommodating immigrants (particularly agricultural workers) from southeastern Anatolia since the nineteenth century (Toksöz 2010). Those families that resettled in the cities are experiencing a culture relatively similar to that in their previous homes. The government also offered a resettlement village, Yalıköy of Didim region, 6 km north of Didim town center on the Aegean Sea coast. Despite the distance, 1,300 km west of their original residence, 110 poor families ventured to move there.

Interviews were conducted using questionnaire sheets, from November 2011 to February 2012, in the resettlement areas close to the Atatürk Dam, including New Samsat, Kahta, and Adıyaman (hereafter referred to collectively as "Adıyaman"). Another interview was conducted with 25 resettlers at Yalıköy in May 2013.

Adıyaman

Ninety-nine resettled families were interviewed and grouped into two classes for analysis based on their income levels. Group 1 consisted of 33 well-off farming families. Each of these farmers were landowners with an income of more than USD 1,000/month, the threshold level for a well-off standard of living at the time of the interview. Group 2 consists of 66 families with an income level below USD 1,000/month. Many of these farmers are seasonal workers, share-croppers, or farmers with land smaller than the country's average size of 5 ha at the time of resettlement (Ballı 2010).

Resettlement and negotiation

Of the 99 resettled families, 55 indicated that details of the resettlement were explained to them beforehand, and 54 indicated that these details were explained either by a government official or the leader of the village. Of the 99 resettled families, only 18 indicated that they had negotiated (or had someone negotiate on their behalf) concerning the resettlement. Only 7 families readily agreed to resettle, while 75 accepted reluctantly, and 17 did not accept the resettlement at all. None of the 99 resettled families was provided land as compensation, but all of them received cash from the government. All of the families within Group 1 had options other than moving to new resettlement areas because of their relatively high compensation;

however, 47 of the families within Group 2 did not. The notification of resettlement was unilateral on the part of the government. The poor families did not seem to have any choice other than to go to the resettlement areas provided by the government, despite the insufficient monetary compensation and lack of a proper resettlement plan.

Farmland

Table 5.5 compares the size of farmland and family income before and after resettlement. While all of the families belonging to Group 1 presently own their farmland, only 51 (83 percent) of the families belonging to Group 2 do. Both the size of the land and incomes decreased after resettlement except for a few large-field owners. Because the income figure is nominal and does not account for inflation throughout this period of time, the income decrease was actually worse than the nominal figure.

Although dissatisfied with the amount of compensation, the large-field owners received a relatively high rate of compensation money from the government, which gave them the chance to select their resettlement sites because they had sufficient funds to establish new farms in many parts of the country. This was not an option for many of the small-field owners due to the low compensation they received for their farm size. Prior to the resettlement, families expected to receive USD 13,000 and USD 25,000 per hectare of cereal land and pistachio-nut orchard, respectively. The actual market price for such land at the time of resettlement varied from USD 10,000 to USD 13,000 (Guler Parlak 2007). However, the compensation actually paid to the families was USD 8,000 and USD 10,000 per hectare, respectively, regardless of the land condition (i.e., depth, slope, stoniness, drainage, texture, etc.) (Guler Parlak 2007). The government compensated the farmers on the basis of property tax statements, which gave much less than the market prices (Demir 2009). In fact, the results of the survey indicated that the actual compensation paid was about 40 percent to 60 percent less than expected. Moreover, the price of land in the newly developed resettlement areas was greater than expected because of land speculation. As a result, the majority of the resettled individuals had to accept less land and

Table 5.5 Median land size and family income

	Land size (ha)		Income (USD/month)	
	Before resettlement	*Time of survey*	*Before resettlement*	*Time of survey*
Group 1	26.0	14.0	2,600	1,900
Group 2	1.4	0.7	500	400

Source: Authors.

less family income than before the resettlement. While Altinbilek and Tortajada (2012) concluded that resettlers had generally been fairly compensated for their loss, they reported a significant delay in resettlement implementation. During the time lag between compensation and resettlement, land prices might have further increased and made it much more difficult for the resettlers to buy new land.

In terms of irrigation, the situation was slightly better. Of the families interviewed, 16 (48 percent) and 50 (76 percent) of Group 1 and Group 2 farmers, respectively, indicated that they presently have sufficient amounts of water for irrigation. The relatively low satisfaction of Group 1 farmers is perhaps due to their experience of sufficient water supply prior to the resettlement. Group 2 farmers with relatively small land were most often close to the banks of the Euphrates River, whereas Group 1 farmers with large fields were often located far from there prior to the resettlement. Following the resettlement, much irrigation infrastructure was developed by the GAP to allow for better irrigation opportunities for large-field owners. Kapur et al. (2009) suggested a double or triple increase in crop yield following installation of the new irrigation infrastructure. Altinbilek and Tortajada (2012) reported the gross agricultural output value of the GAP region after five years of irrigation (in 2000) at around USD 262 million, indicating USD 2,347 per hectare and USD 2,547 per capita. All of this reflects a net increase in income for the region. However, resettled families located in non-irrigated areas of the GAP region were unable to fully enjoy the benefits of this project.

Occupation

While most of the families in Group 1 have continued to farm, more than half of the self-employed farm families in Group 2 changed their occupation (Table 5.6). Eleven resettled families from Group 2 have become laborers, whereas no family was engaged in labor prior to resettlement. This suggests that small landowners could not obtain sufficient land to maintain their lifestyle, and have therefore been forced to change their occupation. To further support this idea, consider that ten of the families in Group 2 indicated that they missed their farmland most desperately after resettlement. As a result, families in Group 2 indicated a general dissatisfaction with their job, an increase from 6 to 49 after resettlement. Twenty (60 percent) and 45 (68 percent) families from Groups 1 and 2, respectively, thought that obesity had increased after resettlement. The amount of self-produced food decreased because they lost farmland or had a smaller piece of land since resettlement. They said that they had to purchase more food from the market and eventually changed their eating habits, consuming fewer vegetables. They felt that the increasing incidence of obesity could be attributed to their changed eating habits.

Resettled families hope that their children will pursue future careers as teachers or government officials (Table 5.7). While 14 resettled families

Table 5.6 Occupations of the resettled families

	Group 1		Group 2	
	Before	*Presently*	*Before*	*Presently*
Self-employed farmer	33	31	44	20
Share-cropper	0	0	9	11
Public sector	0	0	1	2
Private-sector employer	0	0	1	1
Laborer	0	0	0	11
Trader	0	0	0	6
Unemployed	0	0	0	3
Other	0	0	0	12

Source: Authors.

from Group 1 hope their sons will continue farming, only 6 resettled families from Group 2 share this idea. Resettled families from Group 2 are small landowners and some have already abandoned farming; therefore, it is unlikely that they would want their sons to continue farming.

Infrastructure

The quality of life in terms of infrastructure, education, health, transportation, public facilities, and household goods is generally improved since resettlement. Turkey's GDP per capita averaged an annual growth rate of 2.3 percent between 1990 and 2010 (UNDATA 2012), and it is likely that the quality of life of the resettled families improved along with this economic development. Mud and brick houses were common in the homelands of resettled families prior to resettlement; more of them now enjoy living in

Table 5.7 Future occupations desired for children

	Group 1		Group 2	
	Son	*Daughter*	*Son*	*Daughter*
Farmer	14	0	6	0
Company employee	0	0	2	0
Teacher	14	27	19	43
Public official	2	1	31	15
Military service	1	0	4	0
No work	0	5	0	4
Other	0	0	0	0

Source: Authors.

cement houses. However, the average size of a house with a garden decreased from approximately 300 m² (100 m² indoor area) to 200 m² (80 m² indoor area) for families in Group 2. Facilities and opportunities for education have significantly improved, and all of the resettled families indicate that they are happy with this improvement. However, 34 of the families in Group 2 complained that opportunities for employment for their children have worsened. As a result, the majority (89 of 99) of the resettled families feel that the places where they presently live are good in terms of education, while many of them (80 of 99 families) feel that their economic conditions worsened after resettlement.

Local sentiment

A common issue (in addition to the lower income) among resettled families is the fewer opportunities to visit their relatives as a result of their living in different resettlement sites. It is common for those living in this area to frequently visit their parents and relatives. Among the resettled families in Group 2, 65 of the 66 families visited their relatives at least once each month prior to resettlement; after resettlement this number decreased to 40. A similar result was observed among the resettled families in Group 1 (a decrease from 21 to 14). All of the 33 families in Group 1 indicated that losing the opportunity to visit their parents and relatives was very upsetting. Such local sentiment does not seem to be as important among the second and third generations; however, it is an important issue among the first generation.

Resettled families also complained about their loss of social status as a result of resettlement. Owning a farm and a house is a symbol of prestige in southeastern Turkey. Families in Group 1 were mostly landlords of large farms and thus more respected in their homelands. However, the local community of the new settlement sites did not take this social status into account. Families in Group 2 were disappointed that they had lost their farms and were sometimes considered refugees in the new resettlement areas.

Yalıköy

Resettlement

The government's offer for resettlement in Yalıköy Village was accepted by only 110 families. The transportation costs and travel were provided by the government in 1988 and 1989. The government built two-story, 150 m² houses for resettlers, with the first floor intended to be a kind of workshop for income-generating activities for families. A house and farmland (6 ha) were sold to each family at a price of TRY 70 million in 1988, to be paid by 15-year loans and a 5-year deferral on the first payments. This sum was equivalent to USD 38,500 at the time of the loan contract in 1988. The average inflation rate in Turkey in the 1980s and 1990s was over 50 percent; thus, after just five years, in 1993, the typical debt sum had decreased to USD 4,989. The last payments in 2003 were for TRY 4,666,667, equivalent to only USD 3.34. This would be even less than a bus fare from Yalıköy Village to Didim town center. (In 2005, the government redenominated the currency with the old TRY 1,000,000 being equivalent to new TRY 1).

The age of resettlers varied from 60 to 65 years. The resettlers' population at the time of resettlement was around 600 in Yalıköy village and is now around 700. The total number of affected people from the Atatürk Dam living in Didim region is about 1,000, since the resettlers' second generation preferred to move to nearby locations after marriage and employment.

The average education level of the second generation of resettlers is middle school, and that of the third generation is high school, while the children of local Didim residents have mostly attained a university degree. Resettlers reported that following resettlement they focused on economic issues rather than on education; the second generation helped parents in farm work and other jobs. Resettlers interviewed added that because schools providing better education were mainly in the Didim town center, their children's access to better education during the initial years of resettlement was not easy due to the cost of transportation. Second-generation members are still complaining about the low propensity of their children to have a higher education. This may be related to mothers' insufficient capacity to support their children while they attended school, since most second-generation mothers were housewives with a low level of education.

Income

Resettlers generally had more than one income source year-round, both before and after resettlement, such as farming in spring and summer, and trading or working in construction and touristic facilities in autumn and winter. The major occupations and income sources of the resettlers before and after resettlement are shown in Table 5.8. Most of the resettlers were poor before their move. Farming is still the main income source for most in Yalıköy, but the number of share-croppers decreased from seven to two after resettlement. While two of the share-croppers kept the same occupation, four of them became self-employed farmers.

Figure 5.1 compares monthly income of the resettlers in Yalıköy with Group 1 and 2 in Adıyaman, mentioned above. All of the former saw an increase in their income while most of the latter saw a decrease. It is possible to have three crops a year in Yalıköy, but only 1.5 crops in Adıyaman. Olives and vegetables are the main crops in Yalıköy, and they can be sold to touristic facilities, which pay relatively higher prices than town markets. Yalıköy is close to tourist areas, and therefore has more job opportunities in tourism and construction than Adıyaman. Some resettlers live upstairs

Table 5.8 Major occupation and income (USD/month)

	Number		Income (1988)			Income (2013)		
	Before	Present	Max.	Min.	Average	Max.	Min.	Average
Self-employed farmer	9	9	700	320	435	1,700	530	877
Share-cropper	7	2	420	230	312	790	500	623
Public sector	1	0	590	590	590	–	–	–
Private-sector employer	0	1	300	200	260	960	530	745
Private-sector employee	3	5	–	–	–	600	400	483
Laborer	3	3	300	200	253	800	500	710
Trader	2	5	300	300	300	6,000	590	3,573
Unemployed	0	0	–	–	–	–	–	–
Other	0	0	–	–	–	–	–	–

Source: Authors.

in a house built by the government, and rent the downstairs to relatives who arrived later in Yalıköy. Before resettlement, most of the people who resettled in Yalıköy were poorer than those who went to Adıyaman. They were fortunate, however, to be able to purchase a house and farmland with very little payment, and to have the opportunity to find better jobs near tourist areas. Meanwhile, those who resettled in Adıyaman were unable to purchase farmland equivalent to what they had before, due to higher costs resulting from land speculation, and decreased income.

As for the desirable occupation for their children, among the 25 interviewees, 12 and 23, respectively, responded that they did not to want their sons or daughters to continue farming. Among the sons and daughters of the 25 respondents, 15 and 19, respectively, did not want to be a farmer in the future. They have already experienced a better urban life in Didim than they would have had farming.

Social aspects

The original homeland and Yalıköy differ in several respects—from geographic conditions to ethnicity—and the resettlers experienced some problems due to these differences. The first-generation resettlers complained about high humidity in their new location. The fields they cultivated before were more fertile than fields after resettlement due to field slope and stoniness. Wives of first-generation resettlers experienced problems in communication with Turkish-speaking locals, since their mother tongue was Kurdish.

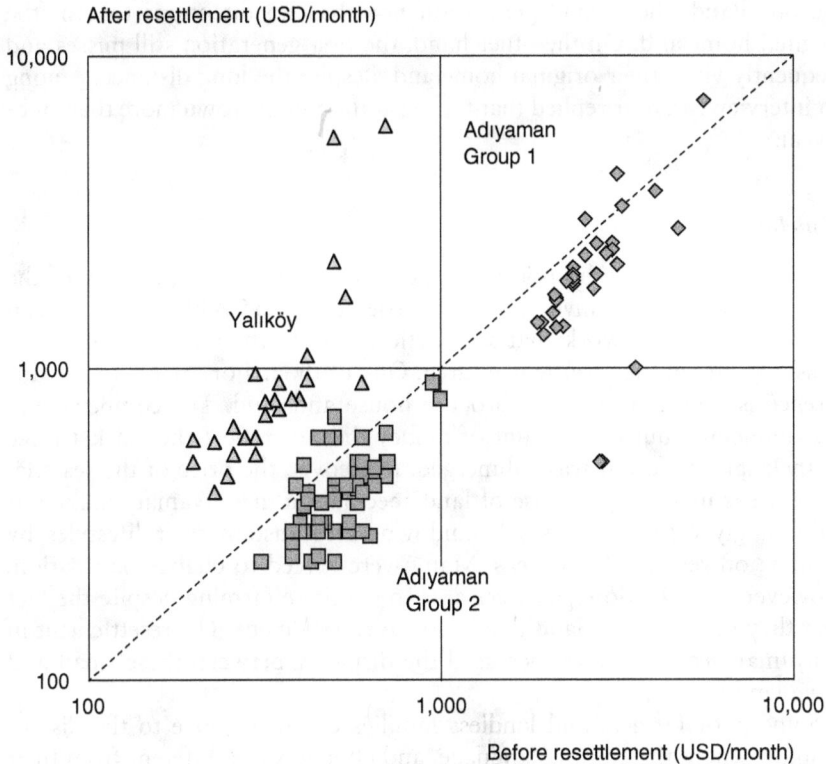

Figure 5.1 Monthly income of resettlers.

Source: Authors.

When Yalıköy village was established 6 km from Didim town center, the resettlers lived in a rather closed society due to their poverty and different ethnicity. In the initial years, cultural differences limited interactions, and locals only established relationships with the resettlers through work. Participants reported that it was ten years after resettlement that social relations began to develop and marriages began to occur between the two communities.

The resettlers accused the authorities of not creating enough opportunities for them to meet with locals during resettlement. As the resettlers' income increased, however, most of second-generation resettlers moved to the Didim town center and relations improved. Also, the second generation is now fluent in Turkish, so communication in society is no longer a problem for them. Thus, relations developed positively, especially through commerce and work. The area's socioeconomic prosperity resulted in a higher level of satisfaction for Yalıköy than Adıyaman resettlers, as shown in Table 5.9. On the one hand, the second generation now has no intentions to visit the original homeland. On the other hand, the first generation still misses and frequently visits their original homeland, despite the long distance. Among 25 interviewees, four replied that they visit their home town more than once a year.

Conclusions

Although our survey did not cover the entire resettlement program of the Atatürk Dam, which involved the resettlement of 55,300 people, we can conclude from our work that satisfaction level of the resettlers depended considerably on the resettlement area. One of the major factors of the difference is how the resettlers procure house and land. The compensation was monetary, but the amount of money did not reflect the market price of their land before it was submerged. Moreover, the price of the resettlement areas increased because of land speculation in Adıyaman, making it increasingly difficult for small landowners to sustain their lifestyles by farming on reduced land areas. Many were forced to change occupation. However, large landowners were able to continue farming despite the fact that they now had less land than prior to resettlement. The resettlement in Adıyaman seems to have increased the disparity between these small and large landowners.

Some poor farmers and landless families dared to move to the distant Yalıköy, where the culture, language, and climate were different from their original home. When each family bought a house and land with a loan, the government maintained the selling prices despite of rampant inflation. As a result, they were able to obtain property at low prices. The resettlement area is located close to tourist areas, so they enjoyed many opportunities for

Table 5.9 Satisfaction with place and job

		Place		Job	
		Before	*After*	*Before*	*After*
Yalıköy	Satisfied	15	22	16	20
	Unsatisfied	10	3	9	5
Group 1	Satisfied	33	2	33	12
	Unsatisfied	0	31	0	18
	No answer	0	0	0	3
Group 2	Satisfied	65	19	60	19
	Unsatisfied	0	45	6	46
	No answer	1	2	0	1

Source: Authors.

employment other than farming, and were able to increase their income. One lesson from this study is that in the case of monetary compensation for resettlement, governments should protect resettlers from the risk of land price increases. Even where inflation is rampant, the selling price of land should be kept as promised, or the amount of compensation should be adjusted upward.

In the Anatolia region, where the Atatürk Dam is located, human relationships among relatives are intimate and emotional attachments to the land are strong, making the emotional impact of resettlement significant. Rich farmers who resettled to Adıyaman complained of the loss of social prestige as landowners. Resettlers in Yalıköy were isolated for some years, until they became accustomed to the different culture and language of the new area. Both in Adıyaman and Yalıköy, the first-generation resettlers missed their home and often visited the reservoir area. For the economically better-off Yalıköy resettlers, however, new prosperity seems to have mitigated their sentiments to some extent.

Miyagase Dam (Japan)

The Miyagase Dam is the second largest dam in the Kanto District of Japan, which includes the national capital region. The construction was implemented as a project under the direct control of the then-Ministry of Construction (MoC) (currently the Ministry of Land, Infrastructure, and Transport—MLIT) at a total construction cost of 399.3 billion yen. The dam was completed in fiscal year 2000, 29 years after the commencement of work. The number of households and population resettled due to the dam are shown in Table 5.10. Because the greatest numbers of dam-affected households are in Kiyokawa, this study follows the history of that particular village.

Table 5.11 shows the history of negotiations relating to the dam construction. Consideration about the dam first began in about 1960. The dam-construction project lasted for about 30 years, from 1969 through

Table 5.10 Area and households submerged by the Miyagase Dam

Name of town	Submerged area (ha)	Displaced households	Displaced population
Kiyokawa Village	374	274	1,104
Tsukui Town	107	1	2
Aikawa Town	9	6	30
Total	490	281	1,136

Note: The town of Tsukui was incorporated into the city of Sagamihara in March 2006.

Source: Kiyokawa Village (1982).

Table 5.11 History of negotiations

Year	Development
1969	Survey on dam construction starts.
1970	Kiyokawa Village and locals establish the Committee for Miyagase Dam Development (CMDD).
1971	MoC establishes Survey Office and requests local communities to consent to a preliminary survey.
1972	CMDD agrees to the survey, expressing ceratin conditions.
1973	Local people's opinions are surveyed. MoC establishes a consultation office.
1974	MoC requests the CMDD to accept a detailed survey on affected people's property. CMDD requests the MoC to submit a Comprehensive Local Development Action Plan as a condition for conducting the detailed survey.
1975	MoC submits the Resettlement Action Plan.
1976	The detailed survey commences. Miyagase Dam Liaison Council (MDLC) established.
1977	A master plan of relocation site development is submitted to the local people. Dam area is designated by law as an area subject to special measures.
1978	Affected people's wishes for livelihood rehabilitation are surveyed.
1979	MoC proposes compensation standards to local communities. Development of resettlement area starts.
1980	A loan facility for rehabilitation of displaced persons' livelihoods and a grant-giving facility for fixed-interest loans to help secure resettlement sites are established.
1981	Affected people agree to livelihood rehabilitation plan proposed by the MoC. Prefectural government proposes "gratitude money." MDLC and Kiyokawa Village agree to the compensation standards. Agreements on resettlement start to be concluded with individual households.
1982	Development of the Miyanosato resettlement area is completed. Actual resettlement begins.
1984	Resettlement to Miyanosato area is completed.
1987	Preparations for dam construction work begins.
1988	All resettlement activities are completed. Actual dam construction begins.
1994	Inundation begins.
1998	Dam operation begins.
2003	All of the special measures based on the ASM are completed.

Sources: Miyagase Dam Construction Office (1992), Kiyokawa Village (1982).

1998, but resettlement-related activities such as compensation negotiations and the establishment of resettlement sites lasted until 1998, 20 years after the announcement of the dam project.

It was in September 1969 that the residents of Kiyokawa first heard an official explanation of the project from the MoC and the perspective of the residents to be relocated. In August 1970, the local residents established the Committee for Miyagase Dam Development (CMDD). In November 1972, an inquiry letter was issued from the local residents regarding compensation and livelihood rehabilitation, to which the MoC replied. Consequently, the CMDD consented to the survey based on two conditions: that the CMDD would be allowed to participate in the survey, and that even after the survey was completed no construction would be done until the local residents approved. In 1973, the CMDD broke up because some groups felt that it was too compliant in negotiations. In 1976, the Miyagase Dam Liaison Council (MDLC) was created as a residents' organization unaffiliated with the village administrative office, and it set out to negotiate with the dam developers. Besides this, a Dam Project Opposition Committee was also established.

In June 1974, the MoC explained the results of the local preliminary survey to local residents and requested approval to conduct a detailed survey to confirm the lot numbers, land-use, and property lines of land and houses, etc., as well as ownership. As one condition to accept the detailed survey, the MDLC demanded that a Development Action Plan be presented relating to the Miyagase Dam construction. This demand was raised, as locals noticed that the government of Kanagawa Prefecture had presented such plans in the case of the Shiroyama Dam and Miho Dam, constructed in the same prefecture, and they demanded the same treatment. The Development Action Plan was very detailed, and besides items for the national government to handle directly, there were also many items for the prefectural government to deal with. After agreement was reached between the prefectural and national governments, a response was issued to the MDLC. The prefectural government also requested cooperation from the local village authorities for the dam construction.

Stakeholders

The stakeholders involved in resettlement of displaced households included the national government ministry in charge of the construction (i.e., MoC), the Kanagawa Water Supply Authority (beneficiary of the water from the dam), local residents' groups (MDLC), municipalities in the watershed (Atsugi City, Kiyokawa Village, Tsukui Town, and Aikawa Town), and agricultural cooperatives. Although the roles of the stakeholders overlapped somewhat, the national government was mainly responsible for establishing the frameworks for compensation of individuals and public bodies.

From 1969 until the present, Kiyokawa Village has allocated two personnel to the Dam Issues Office established within the village administrative office, and it has focused mainly on coordination relating to livelihood rehabilitation issues. One of these two personnel is a resident displaced by the Miyagase Dam. The village office carried out its coordination activities while remembering the emotional or psychological concerns of displaced residents, for example, by spending 20 million yen over three years to move large Himalayan cedar trees that had been growing beside the Miyagase Middle School, and assisted with individual job placements.

The MDLC consists of a secretariat, an expert committee, local committees, and a board of directors, and has four full-time staff. Not surprisingly, many of the residents' views are in conflict with the positions of the national and prefectural governments; thus, so that it would not appear that the MDLC was under undue influence from the government, full-time staff of the Council were seconded from the agricultural cooperative in the city of Atsugi (until mid-1982). In addition, the village office granted about ten million yen per year to the MDLC in the form of subcontracting fees. To provide information and build consensus, the MDLC issued a newsletter starting in 1976 (42 issues in total). It also became involved in establishing standards for loss compensation, allocation of resettlement land, and determination of land-use classifications, etc.

Resettlement areas

Besides the cash compensation provided to the displaced residents based on the National Guidelines (see Box 2.1), assistance was also provided to help them find group resettlement areas in three locations. When polled, many residents in the planned submergence zone (particularly those who were already commuting to Atsugi for work before resettlement) indicated that they desired to resettle away from the village. Thus, in cooperation with Kanagawa Prefecture and Atsugi City, the then-Housing and Urban Development Corporation (a national body) established a property in Atsugi City consisting of about 200 lots, which became the "Miyanosato Resettlement Area," in an "urbanization control area," which was to be kept undeveloped until the municipal government revokes the designation according to the City Planning Act. Lots ranged from 160 to 800 m² in area, considerably larger than in neighboring residential areas. For residents who preferred not to move far from the submergence zone, or those who desired to operate tourism-related shops, construction was commenced around the dam reservoir in Resettlement Area A (general housing) and Resettlement Area B (shops, etc.).

It is important to mention that work had already begun on the resettlement areas prior to the discussions on standards for compensation. The fact that people could see work underway in the resettlement areas, even as negotiations continued, probably reassured the people who were to be displaced, and this likely facilitated the signing of individual agreements after the standards for compensation were settled (Yamaguchi 2003). The allocation of resettlement areas was handled through the MDLC, and the allocation of housing was settled by the MDLC by lottery, after considering the requests of the people affected.

All the land value of the lots in the resettlement areas was about 50,000 yen per square meter. Resettlers were prohibited from reselling these properties for ten years, but market prices per square meter in Miyanosato and Resettlement Areas A and B were about 80,000 yen and 60,000 yen, respectively. It was clear that they were able to obtain land at relatively inexpensive prices (Miyagase Dam Construction Office 1988). Residents who desired to be resettled outside the designated resettlement areas were assisted individually by the prefecture. There were only four such households, and the total land area involved amounted to 1,500 m².

"Gratitude money" and special measures

Kanagawa Prefecture, the beneficiary of the dam project, paid "gratitude money" to each household of local residents from the submerged area. Because the National Guidelines do not recognize compensation for psychological losses, this was a special financial measure, separate from the existing regulatory framework. The amount of the payment depended on length of residence, age of individual, and amount of land area, etc., and a certain assessment was also made for persons renting land or housing. Out of 288 families that received the money, 136 families received between 9 and 10 million yen (Miyagase Dam Construction Office 1992). The total gratitude money paid amounted to 2.8 billion yen.

Kanagawa Prefecture also provided financing and interest assistance for funds needed to obtain new housing or to move, or to secure alternative sites for agricultural or commercial purposes. The prefecture would bear the cost of any interest rate exceeding 3 percent, for loans of up to 20 million yen to acquire an alternative site. A "self-sufficiency rehabilitation fund" provided interest-free loans of up to 30 million yen for three years. In addition, tax reductions and exemptions were instituted to deal with property acquisition tax at the resettlement sites.

The amounts of gratitude money and funds under the financing programs were determined, reportedly, based on examples from other dams in the vicinity, and from the perspective of also enabling land and housing renters to secure a parcel of land (Miyagase Dam Construction Office 1996). As a result, renters of land and housing were also able to own their own homes by using the gratitude money and financing.

In 1997, the dam was designated as being subject to the ASM (see Box 2.1), and consequently an Upstream Regional Improvement Plan was adopted by the national government, paving the way for infrastructural improvements to aid livelihood rehabilitation. To this were added projects implemented separately by the prefectural and town or village offices, leading to the adoption of the Plan. According to the Kanagawa Water Supply Authority (2003), 85 projects were implemented between 1980 and 2003 based on the Plan. When it was initially approved, the project funds were expected to amount to 35.6 billion yen, but due to the lengthening of the time needed for dam construction, as well as increases in project costs, the total cost of the projects rose to 67.9 billion yen and the period was extended. A large variety of projects was implemented, including slope reinforcement, village roads, waterworks and sewerage works, schools, afforestation, parks, childcare centers, and waste treatment facilities.

Consensus-building

The Kanagawa prefectural government was involved in a number of measures related to rehabilitation, including reduction and exemption of prefectural taxes (property acquisition tax), the establishment of permanent consultation offices, the secondment of consultation personnel, and the payment of gratitude money and "cooperation" money. These measures reportedly had an influence on the final consent by residents for the project.

The national and prefectural governments established consultation offices in each village. At the prefectural consultation office, seven permanent staff provided consultation. The consultation office was later combined with a regional government center, but until 2001, counselors were posted full-time. At its busiest time this office handled about 600 consultations per year, covering a wide range of issues, including overall livelihood issues, job placement, procedures for opening a business, assistance for builders, financing, taxes, registrations, and inheritance issues. At the time of resettlement, counselors visited each household about twice per month. In addition, a special department of the prefectural offices has been providing assistance for revitalization of upstream areas since 1977. At present five persons are assigned to this department.

The national government's survey (construction) office for the Miyagase Dam also conducted consultation for local residents. From the time the office was established in 1971 until construction work began in 1988, employment at the office amounted to a cumulative 1,025 person-years. Of that amount, a cumulative 457 person-years of administrative officials' time was allocated to matters related to compensation and other dealings with residents.

These figures are an indication of the desire of personnel of the national and prefectural governments and local authorities to negotiate with locals regarding compensation. From the first proposal of compensation standards in 1979 until consent was reached in 1981, over a period of two years, formal negotiations about compensation were held as many as 150 times. Besides these efforts, consultations and other interactions with resettlers were conducted numerous times, both formally and informally (Yamaguchi 2003).

Separate negotiations were conducted by the national government personnel and each household on a case-by-case basis, but by that point the detailed survey and related negotiations had been completed, and compensation standards had already been established, so residents were probably already in general agreement with resettlement. The main topics of separate negotiations centered on technical aspects, such as addressing differences in the area of land as listed in the land registry compared to the actual land area as determined in the detailed survey.

Infrastructure projects based on the Upstream Regional Improvement Plan were generally completed in 2003, but consultation work relating to rehabilitation has still continued after that under the auspices of the village administrative office, as well as the prefectural and national governments.

Tourism development

In order to promote environmental conservation and tourism development around Miyagase Lake, the national and prefectural governments and local authorities engaged in various improvements of touristic facilities, including parks, cultural centers (facilities to exhibit the traditional livelihoods of residents before resettlement), moorage for tour boats, parking, and automobile-camping facilities. The operation of these facilities is managed by the Miyagase Dam Regional Business Promotion Organization, and all 36 of its employees who manage the facilities are residents originally from the submergence area.

Before the construction of the dam, the valleys in this region were popular for hiking, and the area attracted many tourists. As a result of tourism developments, tourists visiting the area increased from 860,000 persons in 1989, before the dam was built, to 2.83 million persons in 2004. Visitors to the dam area alone reached 1.3 million persons per year. One major reason for this increase is the good accessibility for tourists, as the area is part of a metropolitan region that boasts a huge population. Estimates by the MoC indicate that the Miyagase Dam generates economic benefits amounting to more than 600 million yen per year (MLIT 2006). It is important to note however, that this figure is not restricted to the upstream area.

Costs of rehabilitation and consensus-building

Here, we define "direct" and "indirect" compensation costs. Direct compensation costs in the cost of dam construction include financial compensation for submerged land and property. Compensation costs to individuals were estimated by the MoC in a 1998 study to be about 24.5 billion yen. They are included in the 399.3 billion yen of construction costs for the Miyagase Dam, but costs for the mitigation of social impacts and consensus-building are not included. The calculation does not include any personnel costs of the survey office, as the employees belonged to the MoC. Costs for items under the

Upstream Regional Stimulation Plan (cost of infrastructure improvements for community restoration, personnel costs to assist consensus-building, career assistance, and inducements for employers) are also not included. All of these costs are calculated here as "indirect" compensation costs.

The largest component of indirect compensation costs was the various projects conducted based on the Upstream Regional Stimulation Plan, amounting to a total of 67.9 billion yen. An amount of 1.52 billion yen was provided to the Miyagase Dam Regional Business Promotion Organization as an endowment. In addition, salary supplements, due to increases in administrative work based on the ASM, were provided from the national government to the local authority, with 85 million yen being paid out in 1982. Resettlement sites were offered at lower than the adjacent land values, and the total difference in costs amounted to 3.3 billion yen. In addition, as stated above, gratitude money was paid to people displaced by the dam.

With regards to personnel costs, from the time the administrative staff established the government's construction office until the beginning of work on the dam, if we assume that the entire amount of labor is allocated to resettlement land negotiations, land purchases, consultation, and livelihood rehabilitation measures, the labor would amount to 457 person-years, which we estimate at 2.4 billion yen. A similar calculation was conducted regarding the personnel costs for five persons each from the prefectural government and village office, as well as for establishment of the prefectural livelihood consultation office. The 10 million yen per year the village office paid as an administrative cost supplement to local organizations, as well as the cost of two persons from the village administrative office to deal with resettlement, and personnel costs for the consultation office, are assumed to have matched the personnel costs supplemented by national government payments to the village authority.

Furthermore, costs relating to tax reduction and exemption, job placement, and vocational training, costs of external organizations such as agricultural cooperatives, compensation for holding national assets, and infrastructure projects in the city, etc., might also have been worth considering, but because reliable data could not be obtained, these have been excluded from the calculations below.

The total amount of the indirect compensation is estimated at 78.6 billion yen (Table 5.12), which is not included in the calculation of dam construction costs. These hidden compensation costs amount to 3.1 times the direct compensation costs of 25.4 billion yen. If these indirect compensation costs are divided by 281, the number of resettled households, the amount comes to about 280 million yen per household, equivalent to 43 years of Japan's average household income (6.5 million yen) in 2004. The direct compensation costs amount to only 7.5 percent of the officially announced total construction costs of 399.3 billion yen, but if indirect compensation costs

Table 5.12 Estimate of indirect compensation costs (billion yen)

Compensation	Cost
Upstream Regional Stimulation Plan	67.9
"Gratitude money" (prefectural)	2.8
Loan interest assistance (prefectural)	0.02
Endowment to Miyagase Dam Regional Business Promotion Organization	1.52
Reduced-price sales of resettlement lands (Miyanosato)	3.0
Reduced-price sales of resettlement lands (Sites A and B)	0.3
Personnel costs for livelihood consultation office (prefectural)	0.2
Personnel costs (prefectural)	0.3
Office personnel costs (national government)	2.4
Personnel cost supplement from national government to village administrative office	0.2
Total	78.64

Source: Authors.

are also included, the total construction costs would amount to 477.9 billion yen, and the total direct and indirect compensation costs of 104.0 billion yen would amount to 21.8 percent of total construction costs.

In the case of the Miyagase Dam, even considering compensation costs as large as these, a cost-benefit analysis would indicate that the construction was appropriate. The MLIT has conducted a cost-benefit analysis relating to the flood control benefits of the Miyagase Dam, and according to this analysis, the benefits of 320.9 billion yen are 2.1 times the construction costs of 152.5 billion yen for the flood control portion of the dam (MLIT 2006). Even if indirect compensation costs of 78.64 billion yen are added to the construction costs, benefits exceed costs by a factor of 1.4.

Consideration of local sentiment

A number of measures were taken to consider the sentiments of the residents displaced by the construction of the Miyagase Dam. Examples include the commemorative hall built to portray life as it was before resettlement, as well as a monument inscribed with the names of the former residents, built on the lakeshore. On a tour boat that operates on the dam reservoir, a tape repeatedly announces to tourists that what they are observing exists because of the cooperation of local residents in the dam construction.

The bodies implementing the project dedicated a large amount of effort and time to communicate with the local residents. In the course of resettlement negotiations, even before consent was obtained, the path for some kind of resolution emerged as the parties got to know each other. One official of the time made the following comment:

About once every three days I would make an unofficial visit to a local home, and, particularly during the first month, I just listened quietly to people's concerns or anything they had to say. It was only after listening to everything that I expressed any views from our side.

(Miyagase Dam Construction Office 1981)

Even after people had been resettled, government and personnel from implementation agencies would frequently visit them. These visits played a significant role in reducing the feelings of victimization of displaced residents. By the fact that the authorities were visiting in person, the displaced residents felt that they were not being forgotten. Even if not all their wishes relating to the resettlement had been fulfilled, just the fact that someone would listen to their complaints helped to reduce their dissatisfaction to some extent.

Evaluation of the resettlement

In 1995, the MoC's Miyagase Dam Construction Office conducted a study on the state of rehabilitation of livelihoods of the displaced residents who had been resettled. The result was that 86 percent of households responding to the survey felt that after resettlement their lives had either "improved" or "not changed" (Table 5.13).

These results are quite favorable. One factor in this result might be the composition of employment of those who were resettled. Among the 417 persons who were resettled, 309 were employed prior to resettlement, of who only 19 persons (6.1 percent of those employed) were engaged in agriculture and forestry. More than half of the employed (186 persons, or 60 percent) were company employees, public servants, or part-time employees before resettlement. Many of them had already been commuting to the nearby city of Atsugi, and moving to the Miyanosato resettlement area in Atsugi made commuting more convenient. Only 20 percent of the total number had changed employers after resettlement. Unlike dams built in remote mountain areas where the displaced are most often engaged in agriculture and forestry

Table 5.13 Survey of living conditions before and after resettlement (percent)

	Households	*Percentage*
Improved	77	38.5
Not changed	95	47.5
Worsened	28	14
Sub-total	200	100
No response	33	17
Total	233	

Source: Miyagase Dam Construction Office (1996).

prior to dam construction, the Miyagase Dam was built near a large city. This meant lower than typical impacts on the livelihoods of residents who lived in the submergence zone.

Only 60 households (26 percent of total) responded to the survey regarding income before and after resettlement, and 46 households (20 percent) replied regarding savings. Of these respondents, 36 households replied that income had increased and 22 replied that savings had done so. Only 10 households replied that income had decreased, and 11 households replied that savings had done so. It is important to note that, although the response rates were low, these results at least suggest that the financial condition of resettled households had not deteriorated. The fact that few people actually changed employers may be one factor here.

As for changes in living conveniences, many households mentioned improvements in public transportation, shopping, and health services, although many households felt that they were paying higher prices (Table 5.14). One factor here is probably that after resettlement the affected people ended up living closer to a city. Generally, one could conclude that living conditions improved after resettlement.

In the case of the Miyagase Dam, the assistance provided has been described as a "successful example of rehabilitation" of livelihoods and "the prefecture's best example" of resettlement associated with a dam project. One might be inclined to conclude, therefore, that the overall assessments by displaced residents themselves were relatively positive. But one must remember that these results came after 20 years of consensus-building and resettlement projects, and that at least 78.6 billion yen had been spent for the indirect compensation costs.

It was in the 1980s that the resettlement project of the Miyagase Dam really began in earnest, and this period overlaps with a time when Japan's economy was booming. It was said that the "Miyagase Dam was paved with banknotes." These conditions made it possible to spend such a large amount on compensation, and one could say that the dam proponents were simply lucky in the case of the Miyagase Dam. It is not clear if it would be

Table 5.14 Changes in living conditions

	Improved	Worsened
Education of children	87	25
Health service	133	9
Shopping	150	10
Public transportation	153	11
Relationship with neighbors	60	70
Amusement	80	33
Prices	21	111

Source: Miyagase Dam Construction Office (1996).

possible to provide a similar level of indirect compensation today under Japan's current economic conditions and the financial status of the national and local governments. Because no standards exist to guide indirect compensation costs, on one hand there is the risk that they could escalate out of control, while there is also the possibility, on the other hand, that they could be squeezed to zero. It is important to properly calculate the indirect compensation costs as a part of the total cost of dam construction, and to decide whether or not to build a dam based on such a calculation of costs and benefits. Indeed, many of the costs of dam development are probably underestimated today.

Also, to this day, even if residents in the case of the Miyagase Dam are satisfied with their current living conditions, they still have not cast aside a sense that they became victims of national policies. "It is not surprising that there is still a sense of victimization among residents in resettlement areas. This feeling will probably not completely disappear during the current generation," said a displaced resident from the submergence zone. This statement suggests that there are limits to what can be accomplished by financial assistance alone. Throughout the process, from the initial negotiations to actual resettlement, and, finally, rehabilitation of living conditions, it is necessary for dam proponents to continue being aware of the sentiments and emotional concerns of the displaced residents. In the case of Japan, it has become common practice for residents in a proposed dam submergence zone to examine previous cases in Japan as soon as the issue of resettlement arises, and for them to exchange information with residents who have already experienced displacement. If, in this process, the already-displaced residents express dissatisfaction about the dam proponents, future dam developments will probably become more difficult. It is possible that low-profile and often intangible emotional support provided to residents who are displaced by dam projects actually has the effect of making future resettlement projects go more smoothly.

Discussion

Strong local sentiment attached to the submerged area has been observed almost everywhere among the first generation of resettlers. The case of Atatürk suggests that the emotional issue of the first generation may never be eliminated, as long as they stay away from their original residences. A question to be asked and answered is if moving people along with their assets to a distant place is the right modality for the first generation.

A resettler who was given farmland in a place far from his original residence may not be obliged to farm the land himself. He may rent his newly received farmland to others for money. If the revenue from renting his farmland is sufficiently large for him to act as a landlord, he may stay in the area close to his original residence. It implies that he re-established his livelihood as a landlord, not as a farmer. Alternatively, the first generation could be encouraged to change jobs so that they can live close to their original residences, not as farmers but as workers in other sectors that are not "bound" to the newly acquired farmland.

Employing the "rent scheme" (discussed in detail in Chapter 2) may also be a solution. Income for a resettler is constantly provided as rent for the land under the water, which he still possesses. Since his livelihood is secured as an owner of the land, he may live wherever he wants, including the vicinity of his land (which has been submerged by the reservoir).

As presented in Chapter 2, only 3.9 percent of the resettlers from the Saguling Dam chose to move far way, and it was difficult to secure enough resettlement area around the reservoir. Another example is the resettlers of the Tokuyama Dam in Japan (see Chapter 6), for which resettlement was completed in 1996. One of the resettlers moved to the city downstream, and built a small house at the lakeside of the reservoir. He invites his friends there and stays on weekends about ten times a year, from spring to autumn, longing for his submerged home.

It should be noted that the nature of emotional issues may differ greatly between the first generation, and the second or third generations. Many cases show that only the first generation had a very strong attachment to their land, while the second or third generation is rather indifferent of the original residence. For example, the second generation of Atatürk Dam resettlers that moved away no longer speaks Kurdish (which is spoken around the reservoir), but speaks only Turkish. Compensation packages should be designed and implemented taking into consideration the idea that even within a single family not all individuals may share the same emotion with regard to the place they originally lived.

As indicated by the case of the Miyagase Dam, it might be impossible to completely mitigate the resettlers' local sentiments of loss regarding their submerged homes, however generous the compensation provided. However, if life after resettlement is well rehabilitated, or if, at least, the resettlers had consented to the move, the sentiments of loss may be mitigated to some extent. There seems to be no silver bullet for these issues, but the meaningful participation of resettlers in program formulation and their consent to resettlement appear to be prerequisites. Meanwhile, significant improvements in educational opportunities for the second and third generations may help comfort the first generation, as presented in Chapter 2.

Some resettlers may feel victimized, as seen in the cases of the Atatürk Dam or the Miyagase Dam, even when they have moved to a new place with higher income. However, the sense of having "lost something" in the mind of a resettler may be mitigated by the sense of owning something new, at least to some extent. The provision of opportunities for better education, health care, and welfare for the children of resettlers may work in this context.

Some "added value" could be instrumental to alleviate or eliminate the sense of being victimized. The cases examined suggest that ensuring that the children of resettlers benefit from that "added value" may be a promising way of addressing the emotional issues of some resettlers.

References

Akyürek, G. (2005). Impact of Atatürk Dam on social and environmental aspects of the Southeastern Anatolia Project. PhD thesis. Ankara, Turkey: Middle East Technical University.

Altinbilek, D. and Tortajada, C. (2012). The Atatürk Dam in the context of the Southeastern Anatolia (GAP) Project. In Tortajada, C., Altinbilek, D., and Biswas, A. K. (eds.) *Impacts of Large Dams: A Global Assessment*. Berlin: Springer, pp. 171–199.

Ballı, B. (2010). The landownership in Turkey. *Turkish Agriculture Journal*, 192, 23–29.

Demir, M. (2009). Factors affecting tax evasion. *Journal of Justice, Economy and Social Sciences*. Retrieved from http://e-akademi.org/makaleler/mdemir-3.htm.

Guler Parlak, Z. (2007). *Dance of Life with Water: Dams and Sustainable Development*. Ankara: Turhan.

Kadirbeyoğlu, Z. (2009). *Case Study Turkey: Environmental Change and Forced Migration Scenarios*. EACH-FOR Newsletter, No. 5, EU Sixth Framework Project.

Kanagawa Water Supply Authority (2003). *Fiscal 2003 Summary of Operations*. Yokohama: Kanagawa Water Supply Authority.

Kapur, S., Kapur, B., Akça, E., Eswaran, H., and Aydın, M. (2009). A research strategy to secure energy, water, and food via developing sustainable land and water management in Turkey. In Brauch, H. G., Spring, U. O., Grin, J., Mesjasz, C., Kameri-Mbote, P., Behera, C. N., and Krummenacher, H. (eds.) *Hexagon Series on Human and Environmental Security and Peace* (Vol. 4). Berlin: Springer, pp. 509–518.

Kiyokawa Village (1982). *History of Miyagase Dam Development*. Kiyokawa, Japan: Kiyokawa Village.

Miyagase Dam Construction Office (MoC) (1981). *Ten Years' History of Miyagase Dam*. Tokyo: MoC.

Miyagase Dam Construction Office (MoC) (1988). *Report of the Survey on Regional Development and Life Reconstruction in Myagase Dam Development Area in 1988*. Tokyo: MoC.

Miyagase Dam Construction Office (MoC) (1992). *Twenty Years' History of Miyagase Dam*. Tokyo: MoC.

Miyagase Dam Construction Office (MoC) (1996). *Report of the Survey on Rehabilitation of Living Conditions of Resettlers from Submerged Area of Miyagase Dam*. Tokyo: MoC.

MLIT, Kanto District Regional Improvement Division (2006). Fourteenth Kanto District Dam Management Fellowship Committee. Reference documents for post-evaluation of Miyagase Dam construction project, January 2006. Tokyo: MLIT.

PPLH Unhas (Pusat Penelitian Lingkungan Hidup Universitas Hasanuddin) (1998). *Annual Environmental Monitoring Report.* Makassar: Directorate General of Water Resources Development, Ministry of Public Works, Government of the Republic of Indonesia.

Toksöz, M. (2010). *Nomads, Migrants and Cotton in the Eastern Mediterranean: The Making of the Adana–Mersin Region 1850–1908.* Leiden: Brill.

UNDATA (2012). *GDP per Capita Average Annual Growth Rate.* Retrieved from http://data.un.org/Data.aspx?d=SOWC&f=inID%3A93.

Yamaguchi, S. (2003). *Report on Local Revitalization of Kiyokawa Village.* Text of Japan Dam Foundation Fiftieth Upstream Issues Practices Lecture Series. Tokyo: Japan Dam Foundation, pp. 55–63.

Yoshida, H., Shirai, S., Yamazaki, Y., Suda, M., Doi, N., Shimomura, Y., and Fujikura, R. (2010). Indonesia Bili-Bili Dam Itenjumin no Kurashi ni Kansuru Ichikosatsu [Living conditions of resettlers from submerged area of Bili-Bili Dam in Indonesia: an analysis with sustainable livelihoods approach]. *Ningen Kankyo Ronshu,* 10 (2), 75–90.

Box 5.1 Lao PDR: an Asian battery, meeting the electricity demand of neighboring countries

Some mountainous Asian countries are rich in water resources for hydropower generation. Bhutan, Lao PDR, Nepal, and Myanmar belong to this category. Bhutan, for example, has unexploited hydropower potential of more than 23,000 MW. Bhutan plans to construct six new hydropower stations with a total capacity of 4,484 MW by 2024 (World Bank 2008). Nepal's unexploited hydropower potential exceeds 43,000 MW. Myanmar also has about 39,000 MW of unexploited hydropower potential (World Bank 2008).

Of these nations, Lao PDR is the most aggressive in developing hydropower stations to sell electricity to other countries in the Mekong region. Of the 23,000 MW of exploitable potential hydropower in Lao PDR, about 15,000 MW is internal to the country, and the remaining 8,000 MW represents the country's share in the mainstream flow of the Mekong River, shared jointly with one or more riparian countries (World Bank 2010).

Lao PDR is, relatively speaking, a resource-poor country, except for hydropower potential. Thus, one viable option for the country to secure hard currency is to develop hydropower stations, for the purpose of selling electricity to other countries. Commerce Minister Nam Viyaketh of Lao PDR argued in 2010 that building the energy sector is a key element of the government's plan to reduce rural poverty. At the time, Lao PDR was ranked 133 out 177 nations on the United Nation's

(continued)

(continued)

Human Development Index, which measures education, income, and life expectancy. The World Food Program also estimated that 40 percent of children under the age of five are chronically malnourished (Ferrie 2010).

Meanwhile, Thailand is experiencing a rapid increase in demand for electricity. The annual average growth rate of electricity consumption was 7.1 percent between 1991 and 2010, which resulted in annual electricity consumption of 150,000 GWh in 2010. The annual growth of electricity consumption between 2010 and 2030 is predicted to continue at a fairly high rate of 4.4 percent. Annual electricity consumption in 2030 is predicted to be as high as 350,000 GWh, more than twice the level in 2010 (Fungtammasan and Coovattanachai 2011).

Thailand has plans to import electricity from neighboring countries to meet increasing demand, and has signed memoranda of understanding (MoUs) with five countries (Table 5.15).

Table 5.16 shows Thailand's planned electricity imports. This plan will be implemented in accordance with the country's power import policy, which stipulates that the import of electricity should not exceed 15 percent of the country's total capacity annually (Sirikum 2012).

Table 5.15 Thailand's power purchases from neighboring countries

Country	Capacity indicated in MoU (MW)	Date of MoU
Myanmar	1,500	July 4, 1997
China	3,000	November 12, 1998
Cambodia	Not specified	February 3, 2000
Malaysia	300	May 6, 2004
Lao PDR	7,000	December 22, 2007

Source: Sirikum (2012).

Table 5.16 Thailand's long-term plans for power imports

Power imports	Capacity (MW)
Existing (2011)	2,185
Planned (2012–2020)	3,572
Planned (2021–2030)	3,000
Total	8,757

Source: Sirikum (2012).

Table 5.17 Thailand's planned power purchases from Lao PDR (2012–2019)

Project (power station)	Type of generation	Capacity (MW)	Year of completion
Theun-Hinboun plus expansion	Hydro	440	1998
Houay Ho	Hydro	126	1999
Nam Thuan 2	Hydro	948	2011
Nam Ngum 2	Hydro	597	2015–2016
Hong Sa	Thermal (lignite)	1,473	2019
Xayaburi	Hydro	1,220	2018
Xe-Pian Xe-Namnoy	Hydro	354	2018
Nam Ngiap 1	Hydro	269	2018
Total		5,427	

Source: Sirikum (2012).

Lao PDR presently sells electricity only to Thailand, and has been doing so since 1971, following completion of the 150 MW NN1 Dam. In 1998, the 210 MW Theun-Hinboun Dam began selling power to Thailand as well (Greacen and Palettu 2007). Lao PDR now also sells 1,000 MW of electricity, produced by the Nam Theum 2 Dam, to Thailand (Souksavath and Nakayama 2013). As shown in Table 5.17, Lao PDR's future plans for development of hydropower are integrated into Thailand's long-term plans to import electricity from neighboring countries (Sirikum 2012).

(Mikiyasu Nakayama)

References

Ferrie, J. (2010). Laos turns to hydropower to be "Asia's battery.", *Christian Science Monitor*, 2 July 2010. Retrieved from www.csmonitor.com/World/Asia-Pacific/2010/0702/Laos-turns-to-hydropower-to-be-Asia-s-battery.

Fungtammasan, B. and Coovattanachai, N. (2011). *Thailand's Energy Situation and Major Challenges*. Retrieved from http://stscholar.nstda.or.th/stscholar/csts/images/PDF/sem20130117/file01.pdf.

Greacen, C. and Palettu, A. (2007). Electricity sector planning and hydropower in the Mekong Region. In Lebel, L., Dore, J., Daniel, R., and Koma, Y. S. (eds.) *Democratizing Water Governance in the Mekong Region*. Chiang Mai: Mekong Press. Retrieved from www.palangthai.org/docs/ElectricitySectorPlanning&HydropowerInMekongFull.pdf.

(continued)

(continued)

Sirikum, J. (2012). The Lao's Power Generation from the Thai Perspective. Second East Asia Summit Energy Efficiency Conference. Phnom Penh, Cambodia, July 31.

Souksavath, B. and Nakayama, M. (2013). Reconstruction of the livelihood of resettlers from the Nam Theun 2 Hydropower Project in Laos. *International Journal of Water Resources Development*, 29 (1), 71–86.

World Bank (2008). *Potential and Prospects for Regional Energy Trade in the South Asia Region, Energy Sector Management Assistance Program and the South Asia Regional Cooperation Program*, Retrieved from http://sitere sources.worldbank.org/SOUTHASIAEXT/Resources/223546-11924131 40459/4281804-1192413178157/4281806-1194474073434/SAR_ Energy_Trade_Nov_07.pdf.

World Bank (2010). *Tapping East Asia and Pacific's Hydropower Potential*. Retrieved from http://go.worldbank.org/1TGPVE6VX0.

6 Occupational change from farming to non-farming sectors

Introduction

Safeguard policies of donor agencies, including the World Bank's operational policy (OP4.12) mostly give preference to land-for-land compensation for displaced persons whose livelihoods are land-based. However, there are cases where it is not necessarily appropriate for the policies to be implemented literally.

The World Bank's policy is based on evaluations of large dam projects constructed in the 1980s and evaluated in the 1990s, when resettlers were more isolated from the rest of society than today. Especially where they were ethnic minorities and at a time when primary education had not penetrated into the community, their risk of impoverishment was high unless they received similar alternative land to maintain their original community and way of life. This scenario is not always applicable, however, to large development projects in different socioeconomic contexts. In many Asian countries where universal primary education prepares people to join economic and social activities outside of their local community, most resettlers are able to adjust to the larger society in the country, while some cannot adjust to new environments. In such a context, resettlers should have choices of where to live and how to reconstruct their livelihoods.

In the current era of globalization, the socioeconomic structure in developing countries is changing at a dizzying pace, while life expectancy has been greatly extended. Trying to "preserve" indigenous communities by providing land with conditions similar to their original situation could force resettlers stay in a stagnant situation for the rest of their lives, which could still be many years after resettlement.

Moreover, dams tend to be constructed deep in the mountains. Farmers in such localities often do not possess enough farmland to make a livelihood. For them, land-for-land may not be the best option as a resettlement package, for the farmland to be given after relocation may not be enough to restore their livelihood, so they may need to find additional sources of income. Before resettlement, they may not necessarily have been engaged in farming by their own choice. They may have mostly worked as part-time farmers simply because they inherited the farmland from their parents.

Thus, they may not have been so closely connected to farming, as compared with those who engage full-time in rice cultivation at lower elevations.

Under such circumstances, resettlers sometimes tend to wish to change their occupation to something else, without the need to work as part-time farmers. Moreover, rice cropping or farming in the mountains is a hard job, due to the steep slopes for cultivation and associated difficulties (Interviewees 2014). Resettlers from such settings may not be so keen to restore their livelihood as farmers after resettlement. Deep in the mountains, many people who may be forced to resettle are engaged in forestry. This too is a hard and dangerous occupation. While some may wish to reestablish their livelihoods in forestry, past cases in Japan have shown that many are inclined to find jobs in sectors other than forestry after relocation.

Those resettlers generally hope to move to cities rather than relocating to nearby areas to reestablish their livelihoods in farming or forestry. For example, none of the second generation of NT2 resettlers wished to be a farmer or fisherman in the future. Only 6 of 62 resettled poor farmers from the submergence zone of the Atatürk Dam wished their sons to be farmers, and none of them wished their daughters to do so. A question to be asked and answered is how best they can change their occupation after relocation into cities. They do not necessarily have training and experience to work as specialized professionals. Occupations appropriate for their careers and backgrounds need to be found.

This chapter examines the cases of the Miboro and Tokuyama Dams in Japan and the Bili-Bili Dam in Indonesia, in which many resettlers moved from their original villages into cities. The ways most (if not all) of them successfully rehabilitated their livelihoods after relocation are addressed.

Miboro Dam (Japan)

Miboro Dam was constructed between 1957 and 1961 on the Shokawa River in the Gifu Prefecture of Japan to create 215 MW of hydropower generation, in order to meet the rapidly increasing demand for electricity in post-war Japan. It was the first large-scale rock-fill dam in Japan, with 370 million tons of water storage capacity (Japan Dam Foundation 2014a).

According to Takeishi (1962), 301 households (1,364 people) were obliged to resettle. Resettlement was carried out mostly between 1956 and 1960. The inundated area of the Shokawa and Shirakawa Villages was in a very mountainous area, where people were generally very conservative vis-à-vis city dwellers. It is noteworthy that 69 percent of resettlers surveyed by Takeishi (1962) moved into Gifu City and other cities, including Tokyo. Of more than 200 households of resettlers who used to live in Shokawa Village, only 8 moved within the same village. In case of Shirakawa Village's resettlers, no household remained within the village. Takeishi (1962) suggests that lack of alternative farmland around the reservoir was a major reason behind this outcome. However, the authors found, through interviews with the first- and second-generation resettlers, that conducting agriculture in a mountainous and hilly area used to be very hard work and that few resettlers wished to continue farming after relocation (Interviewees 2014). The developer of the dam, the Electric Power Development Co. Ltd, intended to assist the resettlers to secure farmland in remote areas such as the shores of Lake Biwa (Shiga Prefecture), but few resettlers showed any interest in this option. No systematic measure was thus taken to let resettlers rehabilitate their livelihoods after relocation by continuing to engage in agriculture.

Table 6.1 shows the changes observed in the occupations of resettlers from Shokawa Village in the case of the Miboro Dam. A relatively small number of resettlers seemed to face difficulties by changing occupations. The number appears to have been small due to several reasons. First, many of the resettlers from the dam selected occupations in the city for which no specific training or knowledge was needed. Several resettlers purchased a bathhouse (public bath) in Seki City, Gifu Prefecture (Hamamoto 2014, p. 161), which could be managed by two people (e.g., a husband-and-wife team) without special training. The wife would sit at the reception to collect fees and the husband would maintain the bath and other facilities "behind the scenes." Many other resettlers bought what were known as "love hotels" in urban areas. These hotels were only for short stays of a few hours. The wife would sit in the front to collect room charges without conversation

Table 6.1 Changes observed in occupation of resettlers from Shokawa Village (Miboro Dam) (no. of families)

	Before resettlement	*After resettlement*
Agriculture	147	54
Timber production	13	12
Construction	8	4
Labor	34	6
Civil servant	10	10
Commerce	6	15
Inn/hotel	7	24
Restaurant	6	11
Bathhouse	0	6
Temple	2	2
Freelance	4	4
Other occupations	3	6
Unemployed	5	16
Deceased after resettlement	0	5
Undetermined	0	17
Unknown	0	32
Total	245	245

Source: Takeishi (1962).

with clients and the husband would carry out room cleaning and other tasks. The majority of such hotels in the Maruyama District of Shibuya Ward in Tokyo (where many such hotels are located) used to be owned by the first-generation resettlers from the Miboro Dam construction.

Second, the resettlers from the Miboro Dam managed to avoid problems associated with differences in language (namely, the dialects spoken in rural areas and the standard language in the cities). Owners of bathhouses or "love hotels" had a very limited need to communicate with customers, as the "business" between the owner and customers was carried out almost silently, by the customer paying a fee at the reception. The resettlers were thus fairly free from any inferiority complex caused by their unique dialect after their relocation to the cities. This problem was more significant in the 1950s and 1960s in Japan, when the country was less homogeneous in language than today, with its proliferation of the standard language by television and other mass media.

Third, since the Miboro Dam was among the first major dams developed by the Electric Power Development Co. Ltd, the compensation package given to the resettlers was generous compared with dams built later. While the exact amount of money for compensation has not been disclosed even today, it was assumed to be enough for resettlers to purchase a bathhouse or small hotel (Interviewees 2014). Furthermore, "love hotels" were regarded as a promising business in Japan in those days, when emancipation from old customs and societal traditions was visible among the young generation. Banks were thus keen to lend them money.

Fourth, people in remote villages had strong mutual ties, so they helped each other find new occupations in cities. For the hotels in the Maruyama, one gentleman from Shokawa initiated his business as owner of a "love hotel," then assisted others from the same village, helping them to purchase hotels without charging commission, and connecting them with others in the district to facilitate their new businesses and livelihoods.

Admittedly, the success story of resettlers from the Miboro Dam might not easily be replicated in other dam-construction projects. The Miboro Dam was planned and built to cope with the rapidly increasing demand for electricity at the time. Japan as a whole was then at the beginning of its economic development, and society observed big changes in customs and traditions, which resulted in high demand for "love hotels" in the city center. Not all developing countries where dams are being constructed will be in a condition similar to Japan of late 1950s and early 1960s.

Some important lessons may still be learned here. Changing occupations does not necessarily require specialized education or training, if the selection of new occupations is strategic. Becoming an owner of a business alleviated the difficulties in moving to a city, compared with working under someone else. It was made possible thanks to the generous compensation package and wise selection of new occupation. The existence of a "champion" in a new location and business may also be instrumental for the success to be replicated by others, for following what the champion did may be easier for the followers than doing everything from scratch. We may find from this case some important lessons applicable for future dam-construction projects in the developing world.

Tokuyama Dam (Japan)

Tokuyama Dam is the largest dam in Japan in terms of the water storage capacity (660 million tons). It was constructed by the Japan Water Agency between 2000 and 2008 as a multi-purpose dam on the Ibi River, a tributary of the Kiso River (Japan Dam Foundation 2014b). An agreement was reached in 1983 between the Water Resources Development Public Corporation (presently the Japan Water Agency) as owner of the dam and residents in the villages to be inundated. The resettlement was mostly completed by 1989.

The resettlement program for the resettlers from the Tokuyama Dam was significant in that around 70 percent of the resettlers chose the option of group resettlement. Of 466 households, 331 moved into five districts developed by the Japan Water Agency. These five districts received 83, 79, 73, 65, and 31 households, respectively (Hasebe 2009) and are separated from the dam site by 20 to 40 km. The rest of the resettlers, namely 135 households, relocated to locations of their own choice. Seventeen households moved to Gifu City, the largest city close to their original domicile.

Interviews with the resettlers revealed the following advantages and disadvantages of the group resettlement scheme (Interviewees 2013).

On the one hand, the biggest advantage is that a family resettled with a group of households does not feel isolated in the new location, for they have neighbors from the inundated areas. The members in the group of the resettled households speak the same dialect, have the same background in culture and tradition, and share the history of their original village. They bring the shrines of their original area with them to their new residential area, so that the religious services for their ancestors may be continued (Interviewees 2013). Possible stresses and mental-health problems associated with resettlement may be avoided or mitigated by group resettlement schemes.

Another major advantage of the group resettlement scheme is that the "cost" incurred by the resettlers is small, both in monetary terms, and time and labor consumed to find a destination. Interviewees (2013) mentioned that the Water Resources Development Public Corporation (owner of the Tokuyama Dam) offered the land plots at highly concessional prices for those who followed the group resettlement scheme. The resettlers were thus able to secure more land for residential and other purposes than anyone looking for land alone.

On the other hand, a group resettlement scheme can be accompanied by some disadvantages. For example, resettlers may face difficulties integrating into the host community. The host community was not necessarily hostile, but friendly toward newcomers. Still, community gatherings tended to be organized separately. Festivals and religious gatherings were also held separately by two groups—the original residents and the resettlers. One resettler admitted that the integration of the two took place not with the first but with the second generation, as young resettlers went to the same school from childhood and made their own new friends (Interviewees 2013).

As is often observed in villages to be inundated by dam construction, in the case of the Tokuyama Dam, the village was divided into two groups of people: those who strongly objected to construction of the dam, to the extent they refused to talk with the Water Resources Development Public Corporation; and those who were not necessarily against the idea of dam construction and subsequent resettlement (Hamamoto 2001). The majority of people who selected the group resettlement scheme belong to the second group. Those in the first group mostly chose their destinations by themselves, because moving to a relocation site with the second group was emotionally not acceptable for them. It should be noted that conflicts among village people before reaching the agreement with the Water Resources Development Public Corporation cast a shadow over the resettlers in many respects, including the selection of destinations for resettlement.

While many households moved from their original villages in the heart of the mountains to the areas fairly close to the cities, not many resettlers experienced the difficulties associated with changing occupations (Interviewees 2013). This outcome was presumably due to the following reasons:

1   The working population of the resettled households did not necessarily live deep in the mountains but in nearby cities such as Gifu. Many resettlers were in fact aged and retired persons who did not need to secure occupations in their new domiciles.
2   Those who worked as laborers before resettlement also worked as laborers for construction of the Tokuyama Dam (which lasted ten years or so after resettlement) and other civil engineering works in or close to their original villages. They did not need to find new occupations after relocation.
3   Resettlement took place in the late 1980s, when Japan as a whole was in the midst of an economic boom. Looking for occupations in cities was not difficult in those days (Hamamoto 2001), even for those without any particular training or experience appropriate for the occupations in the cities, such as factory workers (Interviewees 2013).

Table 6.2 shows the changes observed in the number of people per household, before and after resettlement. The number of households with one or two people per household significantly decreased after resettlement. In contrast, after resettlement, more households had four to seven people.

Table 6.2 Changes observed in number of persons per household (Tokuyama Dam)

Persons per household	Before resettlement (%)	After resettlement (%)
1	18.7	8.0
2	35.6	18.6
3	18.7	17.5
4	13.7	18.6
5	8.8	15.5
6	3.4	14.7
7	0.9	5.1
8	0.2	1.4
9	0	0.6
Total (%)	100	100
Year of survey	1984	1992–1993
Number of samples	466	354

Note: The survey after resettlement was conducted on persons who followed the group resettlement scheme.

Source: Tanaka (1994).

Moreover the average age in the resettled households decreased from 45.8 years before resettlement to 41.0 after resettlement (Tanaka 1994). This implies that: (1) before resettlement, more than half of the resettlement households had only a widow, a widower, or a couple of retirement age; (2) many family members in the working generation did not live in the mountain villages but in cities, separately from their parents; and (3) those family members in the working generation joined their parents after resettlement. Interviews with resettlers confirmed these assumptions (Interviewees 2013).

Bili-Bili Dam (Indonesia)

As shown in Chapter 3 and summarized in Table 5.1 (Chapter 5), 2,086 households resettled from the submergence zone of the Bili-Bili Dam ended up in the vicinity of the dam reservoir in the Gowa District (1,079 households, 51.7 percent), urban areas including Makassar and other cities such as Sungguminasa (415 families, 19.9 percent), and transmigration areas (Luwu and Mamuju Districts) (592 households, 28.4 percent).

The authors conducted a series of surveys and interviews with the resettlers from 2005 to 2013. The numbers of interviewees by destination are shown in Table 6.3. Since the relocation was implemented more than ten years ago, there was no updated list of resettlers by category, so it was difficult to select interviewees randomly. The authors therefore asked local authorities to introduce some resettlers, and then asked the interviewees to give names of other resettlers. The authors covered about ten percent

Table 6.3 Number of interviewees

Group	Destination of resettlers	No. of households	No. of interviewees	Year of interviews
1	Reservoir vicinity	1,079	22	2005
			66	2009
2	Urban areas	415	40	2013
3	Luwu District (transmigration)	200	8	2011
			22	2009
4	Mamuju District (transmigration)	392	9	2009
			52	2009
5	Returned to reservoir vicinity (from Luwu and Mamuju)	(Est.) 335	101	2010–2011
	Total		320	

Note: Results of interviews with interviewees other than Group 2 are also shown in Chapter 3 and Chapter 5.

Source: PPLH Unhas (1998).

of each group. Selection bias may have occurred, especially in Group 2, because it is easy to find resettlers who succeed in establishing new life in urban area by using their social networks, while those who are outside of their networks and failed in urban area may be more difficult to find.

For the purpose of evaluating the resettlement to urban areas, the authors examined (1) characteristics of resettlers of each destination group by reviewing data of previous studies, and (2) life changes over the generations by analyzing data for Group 2 collected in the 2013 survey.

Land size in original location and amount of compensation, by destination category

Table 6.4 shows the land size in the original location and the amount of compensation of each group. Resettlers in Group 1 represent affluent households, as land size and compensation ranks in the middle. They could receive enough compensation to find new residences in the reservoir vicinity. Those in Group 2 received the largest amount of compensation and moved to Makassar or other peri-urban areas.

Resettlers belonging to Group 3 and Group 4 had only a small area of land or no land and received the least compensation. They had no choice but to join the TP (see Box 2.2) and kept living there until the time of the interview. In contrast, those in Group 5 had larger areas of land and received more compensation. They joined the TP but later returned to the reservoir vicinity, mostly in the early stages of resettlement. They took advantage of the opportunity to receive land for free. However, they

Table 6.4 Land area in original location and cash compensation received

Group	Land area (ha)		Compensation received (IDR 000s)	
	Min.	*Max.*	*Min.*	*Max.*
1	0.2	5.0	2,500	150,000
2	0.1	11.0	25,000	350,000
3	0	4.0	0	200,000
4	0	2.0	0	100,000
5	1.2	7.0	60,000	200,000

Source: Group 1: PPLH Unhas (2009) (34 respondents at Manuju subdistrict and 32 respondents at Parangloe subdistrict of Gowa District). Group 2: Authors. Group 3: PPLH Unhas (2009). Group 4: PPLH Unhas (2009) (19 respondents at Tobadak Village and 33 respondents at Tommo Village in Mamuju District). Group 5: PPLH Unhas (2011–2012).

preferred to return for various reasons, mainly because they could afford to purchase land near the original location (see Chapter 5).

Ownership ratio of assets

Table 6.5 shows the ownership ratio of assets, by group. It confirms the socioeconomic ranks of resettlers in each group as analyzed in the above section. Groups 1 and 2 seem to enjoy a better life than the others. Group 2 (urban area) was more likely to have assets such as a car and motorcycle

Table 6.5 Ownership ratio of assets

Group	Before relocation (%)					At time of interview (%)				
	1	*2*	*3*	*4*	*5*	*1*	*2*	*3*	*4*	*5*
Color TV	9.1	22.5	13.6	13.5	4.0	93.9	72.5	40.9	28.9	78.0
Black and white TV	18.2	45.0	9.1	9.6	2.0	4.6	27.5	0.0	0.0	16.0
Mobile phone	0.0	2.5	0.0	0.0	2.0	69.7	92.5	72.7	53.8	64.0
Radio	47.0	72.5	40.9	42.3	10.0	54.6	50.0	40.9	46.2	47.0
Bicycle	15.2	52.5	22.7	15.4	9.0	21.2	20.0	50.0	63.5	21.0
Motorcycle	10.6	27.5	18.2	7.7	2.0	65.2	85.0	31.8	48.1	65.0
Car	0.0	12.5	0.0	0.0	–	1.5	17.5	0.0	0.0	–
Refrigerator	1.5	5.0	0.0	0.0	0.0	62.1	87.5	13.6	1.9	50.0
Toilet	10.6	7.5	27.3	26.9	40.0	72.7	97.5	95.5	59.6	80.0

Source: Group 1: PPLH Unhas (2009) (34 respondents at Manuju subdistrict and 32 respondents at Parangloe subdistrict of Gowa District). Group 2: Authors. Group 3: PPLH Unhas (2009). Group 4: PPLH Unhas (2009) (19 respondents at Tobadak Village and 33 respondents at Tommo Village in Mamuju District). Group 5: PPLH Unhas (2011–2012).

before relocation and at the time of the interview. Ownership ratios for some assets differ by local condition after relocation. Ownership ratios for refrigerators in Group 3 (Luwu) and Group 4 (Mamuju) are low because grid electricity was not available.

Reasons to choose the destination

Table 6.6 shows the reasons given by the resettlers of Group 1 and Group 2 regarding their destination after relocation. On the one hand, those who relocated to urban areas were proactive in improving their lives and engaging in non-agricultural work. On the other hand, those who resettled in the reservoir vicinity wanted to live closer to family and continue agriculture, especially rice production. When looking at the reasons not to join the TP, it is obvious that the program was not an attractive option for those in both groups. It should be noted that education was an important factor in the decision to choose urban areas. They purposively choose the destination that has education facilities and they were not sure about the existence of these facilities at the TP area. In fact no secondary school was available close to the TP resettlement areas.

Table 6.6 Reasons to choose the destination (Groups 1 and 2)

Group	Reason to choose the destination	Reasons not to join the TP
1	• Close to family/hometown • Close to family's inherited land • Owning inherited land at the destination • Close to workplace • Good for agriculture/paddy • Want to benefit from the built reservoir	• Reluctant to leave hometown • The TP area is far away • Because of information that floods often occur in TP area • The conditions of TP area are inadequate
2	• Better infrastructure • Close to other family members • Better school • Better health service • Easy to get jobs • Other reasons (owning a business and closer to the market)	• Cannot farm/not strong enough to work • Own a land/housing/and family • Working as a civil officer/have a job/own a business • Was not offered transmigration, because considered as financially capable • Fear of relocating to a distant place • Health conditions • Low education • Children's school

Source: Group 1: PPLH Unhas (2009) (34 respondents at Manuju subdistrict and 32 respondents at Parangloe subdistrict of Gowa District). Group 2: Authors.

Resettlers who joined the TP mostly gave economic reasons: no land remaining (for housing, farming, business); incentives of acquiring two hectares of farmland. When faced with difficulties, many of them returned to the reservoir vicinity and purchased land using compensation money and funds they saved in the transmigration area.

Changes in occupation

Table 6.7 shows changes in occupation. The data indicate that the industrial structure has changed in the reservoir vicinity during the past two decades. The percentage of farmers in Group 1 dropped from 60.7 percent to 38.7 percent, while others (including shop owners, transporters of construction materials) increased to 38.9 percent. In Group 2, the percentage of farmers dropped further, from 67.5 percent to 17.5 percent. However, there is a high ratio of farmers among resettlers joining the TP, especially those in Mamuju, indicating their successful production of cash crops. Among returnees, about 40 percent are still farming, but most of them do not own land, instead working as tenants or laborers.

Levels of satisfaction

Table 6.8 presents resettlers' satisfaction levels and preferences between previous and present conditions. There is no denying that resettlers in all groups enjoy the fruits of economic growth in the country, having more assets than before relocation, so the satisfaction level is generally high. Those belonging to Group 1 and Group 2 show a higher degree of satisfaction than those who joined the TP (Group 3 and Group 4). The richer the family before resettlement, the higher the satisfaction level.

Changes of Group 2 over generations

In this section, the authors analyze changes of lifestyle over the generations using government reports (PPLH Unhas 2013) of the resettlers who moved to urban areas (Group 2). The authors also conducted interviews with 40 households and obtained information on 363 family members (Table 6.9).

This group received relatively a large amount of compensation money and chose to move to urban area for better living conditions and jobs. Table 6.10 confirms their intentions. Among 40 households, only 19 still own farmland and 4 receive an income by selling agricultural products. Obviously most resettlers in this group rely on non-agricultural activities for their livelihoods, and many farming families continue agriculture only for their own consumption.

As Table 6.11 shows, many households have multiple income earners in non-agricultural activities. It is likely that the second generation took over the main income-earner position from parents who retired (Table 6.12).

Table 6.7 Change in occupation (%)

Group	1		2		3		4		5	
	Before	At time of interview	Before	At time of interview	Before	At time of interview	Before	At time of interview	Before	At time of interview
Farmer	60.7	38.7	67.5	17.5	45.5	68.2	87.6	93.2	98	39
Private employee	1.6	3.2	12.5	7.5	9.1	0	1.5	1.5	0	7
Government employee	8.1	9.5	2.5	2.5	9.1	13.6	0	2.6	0	4
Laborer	3.1	1.6	2.5	0	13.6	0	0	0	0	12
Unemployed	15.8	8.1	0	40	18.2	4.5	8.3	0	2	2
Trader	–	–	12.5	27.5	–	–	–	–	0	21
Army	–	–	2.5	2.5	–	–	–	–	–	–
Driver	–	–	0	2.5	–	–	–	–	–	–
Other	10.7	38.9	0	0	4.6	13.6	2.6	2.6	0	15

Note: Where percentages do not add to 100, this is due to rounding.

Source: Group 1: PPLH Unhas (2009) (34 respondents at Manuju subdistrict and 32 respondents at Parangloe subdistrict of Gowa District). Group 2: Authors 2013). Group 3: PPLH Unhas (2009). Group 4: PPLH Unhas (2009) (19 respondents at Tobadak Village and 33 respondents at Tommo Village in Mamuju District). Group 5: PPLH Unhas (2011–2012).

Table 6.8 Satisfaction with current life

Group	1	2	3	4	5
Satisfied with current life (%)	80.2	90.0	77.3	69.9	89.0
Previous condition was better (%)	19.8	10.0	22.7	30.1	11.0

Source: Group 1: PPLH Unhas (2009) (34 respondents at Manuju subdistrict and 32 respondents at Parangloe subdistrict of Gowa District). Group 2: Authors. Group 3: PPLH Unhas (2009). Group 4: PPLH Unhas (2009) (19 respondents at Tobadak Village and 33 respondents at Tommo Village in Mamuju District). Group 5: PPLH Unhas (2011–2012).

Table 6.13 presents subdivisions of occupations of all family members with monthly income, by age. Employees in the formal sector receive a stable minimum income. The income from farming is remarkably smaller than from other occupations. Of female family members in their thirties and

Table 6.9 Age composition by family member category

	Number of people	Age		
		Average	*Min.*	*Max.*
Head of household	40	69.8	35	85
Wife of head	40	66.1	45	89
Child	169	35.8	9	60
Child in law	27	38.9	21	49
Grandchild	77	13.7	1	41
Nephew/niece	3	41.0	35	45
Others	7	88.7	55	110
Total	363	39.5	1	110

Source: Authors.

Table 6.10 Ownership of farmland and income from farmland

	Ownership		Income	
	Before resettlement	*At time of interview*	*Before resettlement*	*At time of interview*
Rice field and dry land	30	2	2	0
Rice field only	4	15	2	3
Dry land only	4	2	28	1
No farmland/no income	2	21	8	36
Total	40	40	40	40

Source: Authors.

Table 6.11 Number of income earners per household

	1	2	3	4	5	7	Total
All households	10	9	8	9	3	1	40
Households whose heads are retired	10	6	3	5	3	1	28

Source: Authors.

Table 6.12 Income earners, by family member category

	Income earner (N)	
	Yes	No
Head of household	12	28
Wife of head	5	35
Child	72	97
Child in law	14	13
Grandchild	6	71
Nephew/niece	0	3
Others	1	6
Total	110	253

Source: Authors.

forties, half are housewives, which means that their households have enough income from the main income earner to feed the wife and children.

The most important factor for resettlers to succeed in establishing non-agricultural livelihoods seems to be the desire to educate their children. Table 6.14 shows education levels, by generation. While 73.1 percent of resettlers in their sixties or older never attended primary school, 89.3 percent of those who were teenagers at the time of relocation and are currently in their thirties completed junior high school, 17.3 percent proceeded to university education, and 42.1 percent of those now in their twenties proceeded to university education. This drastic change in education levels could be attributed to parents' desire to provide a higher education for their children, plus support from the new Indonesian socioeconomic environment.

Conclusion

The resettlement plan for the Miboro Dam was developed in the 1950s, when Japan was about to enter an era of rapid economic growth. The resettlement plan for the Tokuyama Dam was elaborated in the 1980s, when Japan was still

Table 6.13 Occupation and monthly income (IDR 000s) by age

Age	> 60	50–59	40–49	30–39	20–29	12–19	7–11	< 6	Total	Ave.	Min.	Max.
No occupation	52	6	26	22	9	6	2	12	135	–	–	–
Farmer*	1	1	2	5	1	0	0	0	10	72	14	125
Teacher, university lecturer	0	0	8	5	2	0	0	0	15	2,429	1,000	3,500
Government employee	0	2	3	2	0	0	0	0	7	3,056	3,000	3,167
Police, military	0	2	5	2	3	0	0	0	12	2,833	2,700	3,000
Private employee	0	1	2	3	4	0	0	0	10	1,625	1,500	2,000
Entrepreneur	1	0	8	5	0	0	0	0	14	4,625	2,500	10,000
Trader	1	1	2	5	3	0	0	0	12	1,375	1,000	2,000
Real estate	2	0	0	0	0	0	0	0	2	4,542	2,083	7,000
Car/motorcycle service	1	0	2	2	0	0	0	0	5	2,000	500	4,000
Shop/restaurant	1	2	2	0	0	0	0	0	5	1,700	200	5,000
Small vendor	1	2	1	1	0	0	0	0	5	675	200	1,500
Construction	0	2	5	1	0	0	0	0	8	1,840	960	4,000
Machine operator	1	0	1	0	1	0	0	0	3	1,000	500	1,500
Driver	0	0	2	1	0	0	0	0	3	1,500	1,000	2,000
Cattle	0	0	0	1	0	0	0	0	1	20,000	20,000	20,000
Other informal sector	1	0	0	1	1	0	0	0	3	1,050	100	2,000
Student	0	0	0	0	8	33	22	1	64	0	0	0

(continued)

Table 6.13 (continued)

Age	> 60	50–59	40–49	30–39	20–29	12–19	7–11	< 6	Total	Ave.	Min.	Max.
Housewife	4	1	14	20	2	0	0	0	41	0	0	0
Retired**	2	1	0	0	0	0	0	0	3	2,100	2,000	2,300
Unemployed	0	0	0	0	1	0	0	0	1	0	0	0
Hospital/nurse	0	0	0	0	4	0	0	0	4	0	0	0
Total	68	21	83	76	39	39	24	13	363	2,738	100	20,000

Notes: * The authors used income data from the sale of agricultural products in the most recent season. ** Two are former government officers and one is a former banker.

Source: Authors.

Table 6.14 Education level, by age

Age	> 60	50–59	40–49	30–39	20–29	12–19	7–11	< 6
Sample size (persons)	67	21	81	75	38	39	24	13
Never attended school	73.1	14.3	6.2	1.3	0.0	0.0	0.0	100.0
Did not complete primary school	0.0	0.0	1.2	2.7	0.0	0.0	70.8	0.0
Completed primary school	20.9	33.3	12.4	6.7	0.0	5.1	29.2	0.0
Junior high school	0.0	23.8	14.8	26.7	7.9	46.2	0.0	0.0
Senior high school	6.0	19.1	49.4	44.0	47.4	41.0	0.0	0.0
University, college	0.0	9.5	16.1	17.3	42.1	7.7	0.0	0.0
Master's degree and higher	0.0	0.0	0.0	1.3	2.6	0.0	0.0	0.0

Note: Where percentages do not add to 100, this is due to rounding.

Source: Authors.

experiencing an economic boom. Some developing countries including Indonesia are in the same or similar socioeconomic settings as Japan was in the 1950s or 1980s. Some lessons learned from these two Japanese cases should be applicable for dams to be developed in the future in the developing world.

As witnessed in Japan, Indonesia, and other Asian countries, the disparity between farmers and city workers in terms of per capita income tends to widen as the economy grows. This trend, which is probably evident anywhere in the world, may give resettlers of dam projects an inclination to change occupations from agriculture in rural areas to something other than agriculture in cities. The three cases examined in this chapter suggest that securing a job in the city to gain a better livelihood appears to be feasible if some conditions are met.

Selection of an appropriate occupation is one of the major pathways to success, as highlighted in the case of the Miboro Dam. If one looks carefully, some occupations in cities may be suitable even for those who come from rural mountain villages. In countries in the midst of an economic boom, finding a job in the cities may be possible, even for (relatively) untrained persons, as highlighted by the Tokuyama Dam case. This is because the demand for workers is high, to the extent that employers are ready to hire untrained workers and educate them at their own cost. Interviews conducted with the resettlers from the Tokuyama Dam project confirmed this assumption. Many of them successfully changed to new occupations in cities, including employment as factory workers, despite having no appropriate training. On-the-job training was provided by employers.

The Bili-Bili Dam project clearly illustrates the economic advantages enjoyed by resettlers who chose to become urban dwellers, in the same manner as the Japanese cases. Among the 40 households moving to urban areas,

27 household heads were farmers before relocation, but the number dropped to 7 by the time of interviews. The ownership ratio of farmland also halved from 38 to 19 households. Income from agriculture after relocation was much smaller than other non-agricultural livelihoods. It is likely that most households still growing agricultural products are not doing so to sell but for their own consumption. Their main income sources are non-agricultural activities, and they have at least one income earner in each household.

The main income earner in families seems to have shifted to the second generation in the Bili-Bili case, since 40 percent of the first generation has retired. There is a remarkable difference in education level between the first and the second generations; 70 percent of the first generation never attended primary school whereas more than almost 90 percent of the second generation in their twenties completed senior high school and 42 percent graduated from university. Access to better education for children was one of the decisive factors for resettlers to choose the destination of relocation, which was also observed in many Japanese cases, including the Miboro and Tokuyama Dams. This tendency is particularly strong among urban dwellers, and the education level of the family members (in practice, children) was elevated drastically. Many families have more than one income earner from non-agricultural livelihoods. They receive higher income than from agriculture, enjoy better living conditions and more assets. Their satisfaction level is thus higher than in other groups. Their occupation and subsequent income security has been largely improved since resettlement. Even admitting the fact that they were relatively more affluent than others, their choice to shift their livelihood and residences helped them catch up with the rapid socioeconomic growth in South Sulawesi, Indonesia. They took advantage of the monetary compensation scheme of the Bili-Bili Dam project. Provision of generous monetary compensation was also instrumental for the resettlers from the Miboro Dam project to move into cities for better livelihoods than before.

Resettlers who were already relatively well-off before resettlement and obtained much cash compensation, chose to stop farming and moved to urban areas, as seen in Bili-Bili case. While the sampling methodology can be biased, many of them seem to be satisfied with the resettlement. The second generation is able to receive a higher education, unavailable at their original location, as was also the case with resettlers in Japan. They are also able to secure an income from the non-agricultural sector. This substantial improvement of lifestyle would not have been enabled by land-for-land compensation but has been by cash compensation. This is also true with such Japanese cases as Miboro and Tokuyama. The resettlers managed to improve their livelihoods by moving to cities, not by staying in the rural mountains as farmers or foresters. Under certain socioeconomic conditions, as was the case with Bili-Bili, Miboro, and Tokuyama, cash compensation can provide more options than land-for-land compensation.

The socioeconomic setting of the project command area and surrounding regions should be carefully evaluated in terms of existing and future job opportunities, when designing alternatives for resettlers. The three cases examined in this chapter may serve as direction finders for future dam projects in the developing world, where societies may be changing rapidly from having the economy based on a primary industry into one based on secondary and tertiary industries.

References

Interviewees (2013). Interviews with anonymous informants in Gifu Prefecture, September 23, 2013.

Interviewees (2014). Interviews with anonymous informants at Shokawa Town and Gifu City, June 6–7, 2014.

Hamamoto, A. (2001). Reevaluation of the public enterprise and mental damage of the involuntarily resettled locals: a case of Tokuyama dam project, Gifu Prefecture. *Journal of Environmental Sociology* (7), 174–189.

Hamamoto, A. (2014). *Era of Hydropower Station Construction: Memories of Miboro Dam.* Tokyo: Shinsensha.

Hasebe, T. (2009). Rehabilitation of livelihood by appropriate compensation. *Shakaishirinn* 56 (3), 1–29. Retrieved from http://repo.lib.hosei.ac.jp/bitstream/10114/5227/1/56-3hasebe.pdf.

Japan Dam Foundation (2014a). *Miboro Dam, Dams in Japan.* Retrieved from http://damnet.or.jp/cgi-bin/binranA/enAll.cgi?db4=1095.

Japan Dam Foundation (2014b). *Tokuyama Dam, Dams in Japan.* Retrieved from http://damnet.or.jp/cgi-bin/binranA/enAll.cgi?db4=1130.

PPLH Unhas (Pusat Penelitian Lingkungan Hidup Universitas Hasanuddin) (1998, 2009, 2011–2012). *Annual Environmental Monitoring Report.* Makassar: Government of the Republic of Indonesia, Ministry of Public Works, Directorate General of Water Resources Development.

Takeishi, T. (1962). Inundated villages by the Miboro Dam: report of field surveys. *Oita Daigaku Keizai Ronshu,* 13 (4), 1–26.

Tanaka, S. (1994). Change of families due to resettlement by dam construction. *Nippon Jomin Bunka Kiyou,* 17, 117–139.

Box 6.1 Framework for hydropower development in Vietnam and related resettlement policies

Vietnam has had one of the highest economic growth rates among developing countries over the past three decades, with growth occurring at over 6 percent per annum. This growth has been accompanied by a rapid per annum increase in energy consumption, at 14 percent, and installed capacity, at about 12 percent during the 2000s decade (Dinh 2010). Hydropower has also witnessed rapid growth during the

(continued)

(continued)

past three decades, although its share has fluctuated widely, accounting for about 22 percent of total installed capacity in the early 1980s (268 MW of 1,231 MW in 1982), 60 percent in early the 1990s (2,120 MW of 3,510 MW in 1992), and 37 percent in the late 2000s (4,583 MW of 12,357 MW in 2008) (Nguyen 2010). The total theoretical potential of hydropower of Vietnam was estimated at about 35,000 MW, of which the economically feasible potential is estimated at 20,000 MW, with the potential energy production at about 100,000 GWh per annum (Ulfsby 2004). In order to improve the development of energy resources, including hydropower, the government of Vietnam established Electricity of Vietnam (EVN) by Decision No. 562/QD-TTg of the Prime Minister, dated October 10, 1994, on the basis of reorganizing the units of the Department of Energy. EVN became the Electricity of Vietnam Group on June 22, 2006 to promote energy development, including hydropower. Since 2007, the EVN Group has established its public investment arm to mobilize financial resources for the development of energy projects, including hydropower plants under the National Master Plan for Power Development.

On July 21, 2011, Prime Minister Decision No. 1208/QD-TTg approved the National Master Plan for power development for the 2011–2020 period with a vision to 2030 (also known as Power Master Plan VII), to develop the electricity sector in conjunction with national socioeconomic development strategies, and ensure an adequate supply of electricity for the national economy and society, among other development perspectives. This Master Plan prioritized the development of hydropower resources, especially projects with multiple purposes, such as flood control, water supply, and electricity production; bringing the total capacity of hydroelectric power from 9,200 MW in 2010 to 17,400 MW by 2020; and improving the operational efficiency of the system to have the capacity of energy storage hydropower plants to increase from 1,800 MW in 2020 to 5,700 MW by 2030.

As most of the potential large hydropower sites have been developed or designated, the government has focused its efforts on promoting the development of small- and medium-sized hydropower plants, called for efforts to mobilize capital from all economic sectors and to improve mechanisms for collaborative public–private partnerships. The development of this hydropower potential, however, has faced many problems. Tohoku Electric Power Co. (2010) reported that:

> over the past 10 years of our consultation activities in Vietnam, we have also encountered numerous abandoned projects, but in fact of these there are many projects that could be restarted if the necessary technological support and funding could be provided.

This perspective, together with various issues encountered during the construction of these hydropower projects, led to Cabinet Resolution No. 26/NQ-CP dated July 7, 2012 to ensure that "in the process of approval of hydropower projects it is necessary to ensure that the resettlement, environmental protection, forest protection and safety of hydropower projects meet the requirements. Otherwise, they will not be approved."

In this connection, the government decided that the objectives of resettlement policies are to create the conditions for people to quickly resettle in their new living conditions and economic lives on the basis of potential exploitation of natural resources and labor, to gradually change the economic structure of the resettled communities, enhance their productivity, improve income, material and spiritual lives, and contribute to economic development and protection of the environment. In order to achieve these objectives, it would be necessary to ensure effective and good implementation of compensation policies, migration of people, resettlement of households, and the rebuilding of infrastructure as part of detailed resettlement plans (Vietnamese Government 2011). Resettlement plans must be implemented by authorities at all levels, in close cooperation with community-based organizations, according to the following procedures: central government is responsible for regulations, guiding mechanisms, and common policies; and local authorities establish specific details and organizational arrangements for implementation. Detailed resettlement plans are to include three types of resettlements: (1) concentrated rural resettlement areas; (2) urban resettlement areas; and (3) resettlement into existing communities. For each of the three types of resettlement areas, detailed measures are to be drawn up to ensure achievement of the resettlement objectives. For rural resettlement areas, the plans are to provide opportunities for resettlers to diversify agricultural production, improve productivity, and to contribute to environmental conservation. For urban resettlement areas, the plans are to ensure good conditions for the development of the services sector, opportunities for economic diversification, and the provision of urban social services. For the last category, the plans are to include detailed measures to support rapid integration of resettlers into existing communities.

(Ti Le-Huu and Chi Cong Nguyen)

References

Dinh, T. P. (2010). *Power Market Developments in Vietnam*. Hanoi: Electricity Regulatory Authority of Vietnam.

Nguyen, H. H. (2010). *Vietnam Hydropower: Current Situation and Development Plan*. Hanoi: Power Engineering Consulting Joint Stock Company No 1.

(continued)

(continued)

Tohoku Electric Power Co. (2010). *Preliminary Study on Small–Medium-sized Hydropower Development under Build–Lease–Transfer Scheme in Vietnam*. Sendai: Tohoku Electric Power Co.

Ulfsby, Ø. (2004). *Hydropower in Vietnam*. Retrieved from www.e-renewables. com/documents/Hydro/Hydropower in Vietnam.pdf.

Vietnamese Government (2011). Prime Minister Decision No. 193/QD-TTg, "Approval of the Resettlement Plan for the Lai Chau Hydropower Project." January 30, 2011.

7 Conclusion

Introduction

Over the course of nine years from fiscal years 2006 to 2014, this research project conducted post-project evaluations of resettlement programs associated with the construction of 17 dams in six countries. With the exception of NT2 and Yali Falls Dams, the resettlement had been completed 20 years ago or more. In the case of the five dams studied in Japan, more than half a century has passed. For the majority, the second generation of resettlers are now the main income earners in households. Previous post-project evaluations were generally conducted just a few years after resettlement completion, but this research project was able to conduct a longer-term evaluation of resettlement. As a result, the following observations became clear that were not necessarily obvious from short-term evaluations.

Formulation of resettlement programs

When preparing a resettlement program, it is advisable to involve residents at the earliest possible stage. If residents are strongly opposed and refuse to sit at the negotiation table, while the dam construction process simply moves ahead, it will become more difficult to rebuild livelihoods and communities. In the cases such as the Sameura and Miboro Dams in Japan, residents' opposition was strong, and they had no involvement in the preparation of a resettlement program. With Sameura, the dam-construction contractor initiated separate negotiations for resettlement, but this resulted in a heated standoff between residents who were in favor versus those who were against the dam. The negative sentiments on both sides have not subsided even in the half-century since resettlement. Also, because the community was not unified in resettlement negotiations, it lost opportunities to receive compensation for public facilities that were submerged by the dam. If things had turned out differently and the community had been involved earlier in the negotiation process, it would have been easier for the community to rebuild after resettlement. In the case of the Kusaki Dam, the community was able to obtain additional compensation for facilities as well as the training necessary to restore livelihoods after resettlement, and this was accomplished by being actively involved in the resettlement negotiations.

Meanwhile, when residents learn their village is going to be inundated, suspicions can arise if they get the impression that their mayor, chief, or other leaders are involved in making a resettlement program, while giving preference to their own personal interests. This was what was observed in the Saguling case (Nakayama 1998). In such a situation, it is important to note that traditional decision-making mechanisms in the village (e.g., with chief as key person) may not function properly. In other words, it is important to consider the possibility that resettlers may prefer to do their own individual negotiations separately, rather than collective negotiations as an entire village, and that it might be necessary to carefully choose a method of conducting resettlement negotiations.

As in the case of NT2, if a dam is to be constructed on a concessionary arrangement with a predetermined deadline, the time available for a resettlement program may be limited. As a result, it might not be possible to offer the options preferred by residents because there is insufficient time for that option even if it were to be chosen by residents (a minority or the majority).

However, if the resettlement program becomes known by early participation by residents in resettlement programs, in some regions there is the risk that people from other areas will come to live in the submergence zone in the hopes of receiving resettlement compensation. For this, it is important to conduct resident surveys at an early stage, and to identify at the earliest possible stage the residents who qualify for being involved in a resettlement program.

Forms of resettlement compensation

Forms of individual compensation for resettlement are generally based on either cash or land. Table 7.1 shows the primary forms of compensation for resettlers in the case studies covered by this research project. For many cases of cash compensation, the dam proponent develops a resettlement site near the submergence zone, and provides land and houses at prices below market rates. Resettlers who wish to transfer there purchase the land and houses using the compensation money. Of course, resettlers who do not wish to move there search for their own relocation sites and move.

Indonesia generally uses cash compensation, but it also has the TP to encourage people to move from the densely populated islands of Java and Bali to other regions (Box 2.2). In the case of the Saguling and the Wonorejo Dams constructed on the island of Java, the Indonesian government combined transmigration with monetary compensation for dam resettlement and tried to move willing resettlers to other islands. Also, resettlers from the Bili-Bili Dam constructed on the island of Sulawesi were able to resettle on other parts of Sulawesi using the TP if they so wished. In the case of the Koto Panjang Dam, entire villages were resettled together, and a number of options were presented to each village, which then decided as a village on

Table 7.1 Compensation for resettlers whose land was submerged

Country	Dam project	Primary form of compensation
Indonesia	Bili-Bili	Cash
	Koto Panjang	Cash and land
	Saguling	Cash
	Wonorejo	Cash
Japan	Ikawa	Land
	Jintsugawa	Rent for submerged land
	Kusaki	Cash
	Miboro	Cash
	Miyagase	Cash
	Sameura	Cash
	Tokuyama	Cash
Lao PDR	NN1	Land
	NT2	Land
Sri Lanka	Kotmale	Land
Turkey	Atatürk	Cash
Vietnam	Song Hin	Land
	Yali Falls	Land

Source: Authors.

the destination. Farmland and houses were offered at the resettlement site, in accordance with the TP. In addition, monetary compensation was paid for farmland and houses that were submerged.

The Kotmale Dam in Sri Lanka is another example of combining the relocation of residents with a resettlement policy conducted under national policy. Also in Sri Lanka, there was an effort for paddy-field development in what was the less populated arid northwest part of the country as part of the Mahaweli development program (Box 3.1), and resettlers from the Kotmale Dam could choose to move to either the vicinity of the dam or to farmland developed under the Mahaweli program.

Japan adopted guidelines (Box 2.1) relating to land compensation in 1967, establishing cash payments as the form of compensation for property submerged by dams. Prior to that, however, resettlement was compensated with land. The Ikawa Dam is one such example. In the case of the Jintsugawa Dam, the residents continued to own the land even after it was submerged, and, to this day, the power company that built the dam continues to pay resettlers (or their descendants who inherited the rights) to rent the land still now submerged by the dam reservoir. This is an example not seen anywhere else in the world.

In the case of the Atatürk Dam in Turkey, monetary compensation was provided but, at the same time, multiple resettlement sites were developed, and residents who wished to do so could purchase the newly development farmland and houses using compensation money.

Meanwhile, because of different social systems in Lao PDR and Vietnam, compensation was not provided for individual households. Resettlers were moved as a village unit to sites developed by the government.

Compensation by land

Often it is farmers who are forced to relocate due to dam construction, so considering the fact that agriculture is a primary source of their income, it would seem natural that compensation should be done by providing land. In the case of monetary compensation, resettlers receive a large amount of money at one time, which runs the risk that they will spend their money excessively on entertainment, big houses, or cars, and not be able to invest in long-term restoration of livelihood. Miboro, Sameura, and Saguling are examples where residents wasted money on entertainment and were not able to rebuild their livelihoods. It should be mentioned, however, that within the scope of our research, cases like that were the exception. A larger problem was soaring land prices due to land speculation, making it impossible for resettlers to purchase the same area of farmland as prior to resettling using their compensation money, resulting in a decrease in revenues after resettlement. The larger the scale of resettlement, the more likely the problem of land speculation is to occur, which was observed in the cases of Saguling and Atatürk.

In view of the problems that can arise with monetary compensation, aid agencies such as the World Bank and the Development Assistance Committee of the OECD have adopted guidelines based on the principle of compensation by offering alternative land. Our research, however, turned up the following types of problems where alternative land is offered as a form of compensation.

The first problem is that in some areas it is not possible to secure enough alternative land. There have been many such cases in dam construction in Indonesia, and these problems were particularly noticeable with the Saguling and Wonorejo Dams built on the densely populated island of Java. The dam proponents had plans to relocate many resettlers away from Java, in conjunction with the TP, but the response from residents was far less than expected. The outcome was that the dam operators could only provide small plots on low-quality land near the dams or else provide monetary compensation, and, in addition, could not guarantee the provision of alternative land. In the case of NT2, most of the resettlers were farmers engaged in shifting cultivation, but the dam proponent provided land with the assumption that they would settle for paddy cultivation after being resettled. The proponent could only offer each household small paddy fields 0.66 ha in size, however. Also, there was no transfer of farming technology, so resettlers attempted shifting cultivation on land that had been intended to be paddy fields, resulting in a decline in productivity. The result is that many residents cannot support themselves today without governmental assistance. Dam operators are temporarily permitting the use of a large area of public forests near the dams. In some cases, productivity of the farmland provided is not as high as before the resettlement. Farmers who resettled near the Kotmale Dam were given land for tea plantations, but productivity was low due to the steepness of the land.

In the case of the Miboro Dam, residents received a proposal to be resettled as a group based on a land compensation scheme. Even though they knew the proposed site, in the end, they opted for resettling based on individual choices rather than desiring to keep the community together. In this case, the residents saw resettlement as an opportunity to escape from a life of hardship in the mountains, and were reluctant to simply rebuild their original livelihoods at the resettlement site. Thus, almost all residents chose to receive monetary compensation, and after resettling were able to make the transition to occupations other than farming.

As these examples illustrate, with compensation by land there is a line of reasoning that residents who want to maintain the same livelihood as before resettlement should be encouraged to do so, but it is important to recognize that not all rural mountain farmers may want to continue farming at their new location. In many cases, compensation with land should be presented as a major option, but it should not be the only one; other options should also be offered, including cash compensation, with acknowledgment of the wishes of the residents.

Monetary compensation

Compensation by land is based on the assumption that resettlers will continue farming after resettlement. However, it is case-by-case whether resettlers will want to continue farming. There is a tendency for poor farmers and second generations onward to prefer not to continue farming. In Japan in the 1960s, the economy was expanding rapidly, and as a result there was a growing income gap between the agricultural sector on one hand and the industrial and service sectors on the other. This situation is similar to what we see in Southeast Asia today. In such economic conditions, land compensation would end up restricting resettlers to farming, which could not offer much expectation of income growth. Cash compensation also provides the chance for resettlers to distance themselves from farming and begin a new life as urban dwellers.

In this study, among the residents who were resettled due to dams in Japan and the Bili-Bili Dam, particularly the second generation and later, there were many residents who stopped farming and succeeded economically. Monetary compensation comes with the risks mentioned above, but if excessive spending can be avoided and residents can be given opportunities and training for new occupations, one can expect better lifestyle improvements after resettlement compared to cases of land compensation. In the case of the

Wonorejo Dam, resettlers desired occupational training. In the case of the Miboro Dam, the occupations chosen by resettlers who had stopped farming and moved elsewhere included the management of public bathhouses and small hotels, which did not require special skills. If the proper occupational training had been offered to these residents, their career options would have been even greater after resettlement.

As in the case of the Miyagase Dam in Japan, there are ways to facilitate the creation of a better foundation for living after resettlement, such as by developing new housing in the resettlement area, providing it at lower than market prices, and providing interest-rate subsidies on housing loans, while also offering cash compensation. These are effective approaches for residents who are interested in keeping their communities together after resettlement.

In a different example, the Atatürk Dam, some of the resettlers borrowed money and purchased land in the resettlement area, and thereafter they were still able to obtain land at a set price despite strong inflation, with the result being that residents were able to buy land at very low prices. In this example, resettlers who attempted to purchase land near the submergence zone suffered losses due to surging land prices, but resettlers who had purchased land developed more remotely were able to obtain enormous profits as they were able to purchase at predetermined prices despite later inflation. When using cash as compensation, some consideration should be given to ensuring the ability to cope with fluctuations in land prices and other prices.

Implementation of resettlement programs

There are some cases of resettlers enduring significant losses due to resettlement operations not going as planned. In the case of the Koto Panjang Dam, the entire village moved with the resettlement, but they suffered major losses in income as the rubber plantations they had been promised were not ready by the time they were resettled. In the case of the Bili-Bili Dam, many resettlers returned later to their original locations near the submergence zone due to many problems, including friction with indigenous people who asserted ownership rights on the resettlement land the resettlers had chosen under the TP, the fact that farmland provided was much smaller than the 2 ha promised, and lack of roads to provide access to markets.

A major cause of problems in implementation is compartmentalization and poor coordination among implementation agencies. For example, in the case of the Koto Panjang Dam, the power company built the dam, but the local government prepared the resettlement site. The power company was motivated to complete the dam as quickly as possible, while the local government had no motivation to prepare the resettlement site quickly. Liaison and coordination meetings were held among agencies involved in resettlement, but in many cases they did not function as intended, and there was a lack of information sharing. Also, although adequate human and financial resources were invested in the design and construction of the dam itself, in many cases, the agencies responsible for resettlement implementation lacked both types of resources.

One way to ensure proper implementation of the resettlement program is to make the start of actual dam construction conditional on progress with the resettlement program. Rather than starting dam construction as soon as residents have been removed from the site, the construction should wait until residents have a good prospect of livelihoods being restored at the resettlement site. In cases where a resettlement site is being developed, dam construction should proceed only on condition that the site has been prepared and resettlement completed. Rather than resettlement of residents being considered one part of the dam development project, it should be implemented as a separate project, and as a condition to be met before implementation of the dam construction. If there is a rush to construct the dam, this means that adequate human and economic resources should first be provided in a concentrated way for the resettlement work.

Also, the resettlement program should not end when residents have been relocated. There is a risk that unexpected events could occur after relocation. In the case of the Bili-Bili Dam, friction arose between the resettlers and indigenous people in the resettlement area, and the resettlers also suffered major setbacks when their orange plantations were hit by disease. In the case of the Ikawa Dam, where the resettlement site was developed based on rice farming, the market did not accept the rice produced there due to its low quality, and changes in national policies for rice growing made it difficult to sustain livelihoods based on rice farming alone. Paddy fields were provided to resettlers in the case of the Kotmale Dam as well, but a drop in international rice prices prevented improvements in farmers' incomes. Resettlement programs should include long-term monitoring of resettlers' living conditions after resettlement, and have the flexibility to take steps to respond with new measures as necessary. In the case of the Miyagase Dam, a system was set up to support the long-term restoration of livelihoods of residents after resettlement, with two full-time staff at the municipal government office assigned to the task.

It is also worth considering accumulating money in a fund from a portion of dam profits (e.g., electricity revenues in the case of a hydropower project), as a method to secure the necessary funds to deal with unexpected events, and for maintenance (e.g., repairing roads that provide access from the resettlement area to urban areas in the vicinity) of the resettlement area after dam construction has been completed (Nakayama 1998).

Supplementary income sources

Income diversification and supplementary income sources are common features in many successful cases of farming continuation and livelihood restoration after resettlement. In the case of the Kotmale Dam, resettlers considered income stability to be as important as income increases. A dependency on one crop could lead to instability of income in the event of fluctuations in crop yields and market prices. It is desirable to design resettlement programs to secure supplementary income sources.

In the case of the Saguling Dam, constructed near a large city, many small shops appeared near the dam reservoir, and were a secondary development to the dam construction. Very few of the shop owners were resettlers, however, but people who had migrated from elsewhere. This displacement of resettlers occurred due to the lack of training for them, and the lack of funding required to open shops (Nakayama 1998).

Considering the reality that many resettlers would prefer to resettle in the vicinity of the dam reservoir, fish aquaculture is one highly feasible option to secure supplementary incomes. There have been successful examples of building aquaculture ponds, as in the case of the Koto Panjang Dam, even (as in Saguling) without using the actual dam reservoir surface. Where aquaculture is conducted in the reservoir, the resettlers may lose out in competition with outsiders unless the resettlers are given some exclusive rights to use the reservoir surface. Furthermore, if farmers have no prior experience with aquaculture, it is important to provide funding for initial investments, and training in aquaculture techniques. This failed in the case of the Saguling Dam, as it was outside "investors" who profited from aquaculture, not the resettlers.

If farming is the only industry that continues after resettlement, crop diversification will contribute to livelihood restoration. A likely cause of differences in income levels between the Yali Falls Dam and Song Hin Dam in Vietnam was a difference in the level of crop diversification.

Educational environment

This research project found that, almost without exception, residents increased the value of their property and were materially better off after resettlement. However, it cannot be said that this increase in property was a function of the resettlement. Rather, a major factor to be considered was that more than 20 years had passed since resettlement, and in that period the national economy had grown. The resettlement effects cannot be analyzed unless a comparison is made with non-resettlers who lived in the same area.

Even so, it is appropriate to assume that the educational environment improved considerably with the resettlement. In almost all cases, the educational situation prior to resettlement was rather poor, and, in many cases, students would have had to commute large distances to get to intermediate and higher levels of schooling. For example, there was no primary school in the original village in the case of the NT2 Dam. The majority of resettlers had received absolutely no school education, or at most had completed primary school.

The educational situation after resettlement clearly improved. Almost without exception, primary and middle schools were constructed in the resettlement zone, so the second generation certainly enjoyed improvements in the level of education. Even resettlers who believed their economic situation had worsened felt that their educational environment had improved after resettlement. Many residents highly appreciate improvements in the educational environment. Resettlers from the Ikawa Dam cannot say that their lives improved economically, but some residents say that the best thing was that the second generation had been able to receive a higher education. Looking at the second generation of residents who resettled to urban areas from the Bili-Bili Dam, 42 percent of young persons in their twenties had obtained university education, and some were even enrolled in graduate school.

Improving the educational environment is one way to increase the satisfaction levels of resettlers. Assistance should be provided in the form of facilities and financial support so that resettlers are able to receive at least intermediate education at the resettlement site. It should also be possible to establish scholarship funds for resettlers from the proceeds of the dam.

Attachment to the land

In many cases, residents have a strong attachment to their original area, and wish to continue living in the vicinity even after the land has been submerged. In the cases of Saguling and Wonorejo, the TP offered land to resettlers to move them away from the island of Java, but the uptake from residents was far less than what the government had expected, with the result being difficulties in securing land for resettlement close to the dam.

Nevertheless, no matter how generous the compensation or successful the livelihood restoration after resettling, the first generation that directly experienced resettlement typically has a strong attachment to the original land, making it difficult to completely assuage the sense of loss. In the case of Japanese dams, the first-generation resettlers still gather regularly on the lakeshore to hold festivals in memory of their submerged home town. At Tokuyama Dam, resettlers have built villas along the lakeshore and in nearby towns, and come to visit regularly even if they have resettled far away. This behavior is also observed with the Kusaki Dam, and the Atatürk Dam. In the case of the Atatürk Dam, the emotional stress is a major factor for resettlers who failed to rebuild their lives after resettlement, and those who lost their status as owners of land there before relocating.

The "rent scheme" used in the case of the Jinstugawa Dam is a unique story, but is worth considering as one option to deal with emotional issues such as these. In that particular case, since the resettlers continue to "own" the land they owned before resettling, to some extent they are able to maintain status as landowners and have a feeling of being "residents" of that land. However, several things need to be considered for "rent schemes." In the case of the Jintsugawa Dam, the dam operator must continue paying rent in perpetuity. An alternative might be a "loan scheme" that limits the land "rental" to 50 years, after which the ownership rights automatically transfer over to the dam operator. From the second generation of resettlers onward, the feeling of attachment to the submerged land may be weaker, and they tend to consider the resettlement area where they grew up to be their home, so one would expect less emotional resistance to a transfer of ownership rights 50 years after the original resettlement.

Evaluation using the IRR Model

Regarding the risks of resettlement on resettlers, Cernea recalls that the IRR Model for resettling displaced populations, which he advocated, has been applied in many actual cases (Cernea 2014). In Cernea (2000), its empirical validation is provided based on seven resettlement cases analyzed by the Institute for Socioeconomic Development in Orissa, India, as well as cases in Lesotho and Nepal. Although the purpose of this research project is not to validate the IRR Model, this section uses the IRR Model as a tool to analyze the findings obtained herein.

The IRR Model lists eight types of "major impoverishment risks" and here we also apply the "differential risk intensities" and "risks to host populations" identified by Cernea (2000) as risk factors. Table 7.2 summarizes the presence of these 10 risks for the 17 cases analyzed in this research project. In the table, major risks that were observed are indicated with an "X," while major risks that we believe were addressed appropriately by "good practices" at the time of resettlement are indicated with a "G."

It must be noted that Table 7.2 is based only upon interviews with a limited number of resettlers, so it does not necessarily cover all possible major risks. For example, even if someone did lose land or work as a result of resettlement, no "X" could be entered in the table for "joblessness" or "landlessness" if such a person was not the actual interviewee or the interviewee did not allude to such a person. Even though the summary does have such limitations, it is still worth considering some of the observations from this table.

The first is that even with dams for which resettlement could not be considered to be a success (e.g., Saguling), no more than three risks were observed. Conversely, even for dams where resettlement is considered to have proceeded relatively smoothly (e.g., NT2, Miyagase), at least one major risk was observed. In other words, one cannot simply say that dams with major risks are examples of failure, and dams with no (or few) major risks are examples of success.

The authors hold the view that, when it comes to formulating a resettlement program, it is difficult to entirely eliminate all of the risks indicated here from the resettlement program, and one or two of the risks are inevitable. There is a practical choice of which resettlement methods to select, considering which of the many risks to avoid and which to "accept" in order to minimize the overall risk. The important thing is not to reduce the number of risk items, but rather, to reduce the overall risk of the entire resettlement program (i.e., minimize the negative impacts on resettlers).

The IRR Model is useful as a checklist when formulating and implementing resettlement programs, by explicitly representing patterns of risk expression

Table 7.2 Major impoverishment risks observed in each case

Country	Dam project	Major impoverishment risks								Differential Risks	
		Landlessness	Joblessness	Homelessness	Marginalization	Food insecurity	Increased morbidity	Loss of access to common property resources	Community disarticulation	risk intensities	Risks to host populations
Indonesia	Bili-Bili	X		X							X
	Koto Panjang	G	X			X			G		
	Saguling	X	X		X						
	Wonorejo	X									X
Japan	Ikawa	X	X								
	Jintsugawa				G						
	Kusaki		G								
	Miboro		X								
	Miyagase	G		G				G			X
	Sameura	X							X		
	Tokuyama	X		X					G		X

(continued)

Table 7.2 (continued)

Country	Dam project	Major impoverishment risks								Differential Risks	
		Land-lessness	Job-lessness	Home-lessness	Marginalization	Food insecurity	Increased morbidity	Loss of access to common property resources	Community dis-articulation	risk intensities	to host populations
Lao PDR	NN1					X					
	NT2					X		G	G		
Sri Lanka	Kotmale		X								
Turkey	Atatürk	X			X						X
Vietnam	Song Hin	X									
	Yali Falls					G					

Notes: X: Major impoverishment risk observed. G: Good practices used appropriately to manage major impoverishment risk.

Source: Authors.

associated with resettlement. It could not, however, be described as a score-board to evaluate whether a resettlement program is good or bad. This is because in many cases the risks indicated here depend on unavoidable factors such as the location and characteristics of a project site.

For example, in cases where it is not possible to provide adequate farmland for all resettlers due to dam site-related factors, an effective option might be to relocate residents to urban areas in order to avoid "joblessness." In that case, in order to avoid "homelessness," as done in the case of the Tokuyama Dam, a suitable option might be to build a collective resettlement site in the urban area and allocate it to resettlers. In this case, however, it may be difficult to avoid the occurrence of "risk to host populations." In the case of the Tokuyama Dam, there is still some "distance" between the resettlers and host population today, even though more than 20 years have passed since resettle-ment. In other words, in cases like this, it is not possible to entirely eliminate three risks: joblessness, homelessness, and risk to host populations.

Also, in order to avoid "landlessness" at the resettlement site for farmers who resettle, an effective option may be to create farmland and housing land from as-yet undeveloped land that had not previously been developed for farming. In the case of Koto Panjang, landlessness was avoided by securing farmland in this way at the resettlement site. Meanwhile, in the case of Saguling, many resettlers experienced "landlessness," and the "homelessness" and "marginalization" that arise from homelessness, because residents refused to leave Java to live on another island. In Kotmale, however, which avoided "landlessness," many people (particularly the second- and third-generation resettlers) experienced joblessness due to the lack of employment opportuni-ties, because the farmland and housing at the resettlement site were in a remote area, far from the city. In the case of Saguling, the dam itself was in the suburban area of the large city of Bandung, so various employment opportuni-ties were more abundant than for Kotmale, even for the first-generation resettlers (though whether or not they were satisfied with the level of income is a different issue). In a situation like this, it is difficult to avoid joblessness for the second and subsequent generations, as well as avoiding landlessness.

References

Cernea, M. M. (2000). Impoverishment risks, risk management, and reconstruction: a model of population displacement and resettlement. Risks and reconstruction: experiences of resettlers and refugees. Paper presented to the UN Symposium on Hydropower and Sustainable Development, Beijing, October 27, 2000. Retrieved from http://communitymining.org/attachments/254_population_resettlement_IRR_MODEL_cernea.pdf.

Cernea, M. M. (2014). Personal communication, September 4, 2014, Washington D.C.

Nakayama, M. (1998). Post-project review on environmental impact assessment methodology applied for Saguling Dam for involuntary resettlement. *International Journal of Water Resources Development*, 14 (2), 217–229.

Index

For Product Safety Concerns and Information please contact our EU
representative GPSR@taylorandfrancis.com
Taylor & Francis Verlag GmbH, Kaufingerstraße 24, 80331 München, Germany

www.ingramcontent.com/pod-product-compliance
Lightning Source LLC
Chambersburg PA
CBHW050421280326
41932CB00013BA/1948

9 781138 597860